# Synthesis Lectures on Technology and Health

The series publishes state-of-the-art short books on transformative technologies for health, wellness, and independent living. Our scope of publishing in the expanding health tech field includes:

- Technology in support of active and healthy living and aging
- Digital technologies for health- and social-care improvement
- Diagnostic, screening, and tracking tools
- Assistive and rehabilitative technologies

The series includes a subseries of books published in partnership with Canada's AGE-WELL that specifically addresses their 8 AgeTech Challenge Areas. Each lecture introduces the context in which the technology is used—wellness, health, medicine, special needs, or other contexts. Authors present and explain the technology and review promising applications and opportunities as well as limitations and challenges. They include material on their own work while surveying the broader landscape of related research, development, and impact.

Walter R. Boot · Andrew Dilanchian ·
Saleh Kalantari · Sara J. Czaja

# Extended Reality Solutions to Support Older Adults

## Potential Applications for Users With and Without Cognitive Impairments

Walter R. Boot
Center on Aging and Behavioral Research
Weill Cornell Medicine
New York, NY, USA

Andrew Dilanchian
Department of Psychology
Florida State University
Tallahassee, USA

Saleh Kalantari
Design and Augmented Intelligence Lab
Cornell University
Ithaca, USA

Sara J. Czaja
Center on Aging and Behavioral Research
Weill Cornell Medicine
New York, NY, USA

ISSN 2771-7054 ISSN 2771-7070 (electronic)
Synthesis Lectures on Technology and Health
ISBN 978-3-031-69219-2 ISBN 978-3-031-69220-8 (eBook)
https://doi.org/10.1007/978-3-031-69220-8



This Springer imprint is published by the registered company Springer Nature Switzerland AG
The registered company address is: Gewerbestrasse 11, 6330 Cham, Switzerland

*We dedicate this book to the older adults in our lives who have inspired us, the many research participants in our studies who have devoted their time and effort to support the advancement of knowledge, and the students and research assistants who have aided in conducting these studies, making this work possible.*

# Acknowledgements

The authors would like to thank the Science and Technology for Aging Research (STAR) Institute, of Simon Fraser University, for its administrative and financial support and the AGE-WELL Network of Centres of Excellence for their financial support. A special thank you to Juliet Neun-Hornick at Simon Fraser University for her administrative and coordinating roles with our team and with Springer.

The authors would particularly like to thank everyone in the AGE-WELL community—Network office, researchers, older adults and caregivers, and partners. Their passion for innovation was essential in developing and sustaining this book's creation. Thank you for your collaborative spirit.

We also gratefully acknowledge support from the National Institute on Disability, Independent Living, and Rehabilitation Research (NIDILRR Grant Number 90REGE0012) under the auspices of the Rehabilitation and Engineering Research Center on Enhancing Neurocognitive Health, Abilities, Networks, and Community Engagement (ENHANCE; www.enhance-rerc.org), and from the National Institute on Aging (P01AG073090) through the Center for Research and Education on Aging and Technology Enhancement (CREATE; https://create-center.org/).

Finally, we would like to express our gratitude to Barry Pendergast for his thoughtful and insightful comments and discussions that have shaped the final version of this book.

AGE-WELL (www.agewell-nce.ca) is Canada's Technology and Aging Network. The pan-Canadian network brings together researchers, older adults, caregivers, partner organizations, and future leaders to accelerate the delivery of technology-based solutions that make a meaningful difference in the lives of Canadians. AGE-WELL researchers are producing technologies, services, policies, and practices that improve quality of life for older adults and caregivers and generate social and economic benefits for Canada. AGE-WELL's work is supported through Government of Canada funding programs.

The STAR (Science and Technology for Aging Research) Institute (www.sfu.ca/starinsti-tute) at Simon Fraser University (SFU) is committed to supporting community-engaged research in the rapidly growing area of technology and aging. The Institute supports the development and implementation of technologies to address many of the health challenges encountered in old age, as well as addresses the social, commercial, and policy aspects of using and accessing technologies. STAR also supports the AGE-WELL network.

# Contents

Contents

# About the Authors

**Walter R. Boot** Ph.D., is the Irving Sherwood Wright Professor in Geriatrics II within the Division of Geriatrics and Palliative Medicine, and Associate Director of the Center on Aging and Behavioral Research at Weill Cornell Medicine. He received his Ph.D. from the University of Illinois at Urbana-Champaign in Cognitive Psychology in 2007. Dr. Boot is one of five principal investigators of the multi-disciplinary Center for Research and Education on Aging and Technology Enhancement (CREATE), a long-standing and award-winning National Institute on Aging-funded center dedicated to ensuring that the benefits of technology can be realized by older adults. He is also Co-director of the Enhancing Neurocognitive Health, Abilities, Networks, and Community Engagement (ENHANCE) Center, funded by the National Institute on Disability, Independent Living, and Rehabilitation Research, with a focus on how technology can support older adults living with cognitive impairments. His research interests include how existing and emerging technologies, including virtual reality and artificial intelligence, can support the health, wellbeing, quality of life, and social connectivity of older people. He is a Fellow of the American Psychological Association (APA) and the Gerontological Society of America.

**Andrew Dilanchian** Ph.D., holds a doctoral degree in Cognitive Psychology from Florida State University, specializing in human factors as it relates to virtual reality (VR) programs. His research focuses on how VR influences presence, immersion, and embodiment, as well as how VR may be used for the assessment and detection of cognitive impairments. Currently, he is working as a post-doctoral research fellow at the Army Research Institute, where he continues to explore the intersection of cognitive psychology and VR technology.

**Saleh Kalantari** Ph.D., is an Associate Professor in Cornell University's Department of Human Centered Design. He is the Director of the Design and Augmented Intelligence Lab (DAIL) at Cornell, where his research group explores human-technology partnerships in the design process, creating opportunities for innovation and creativity. His work advances empirically grounded design, developing tools and techniques to improve understanding of design's impact—both virtual and built—on human behavior. Dr. Kalantari was awarded a National Science Foundation CAREER Award for his research agenda. His translational research has also earned recognition, including a Touchstone Gold Medal Award from the Center for Health Design and a nomination for the National Design Award by Cooper Hewitt. His work is supported by the NIH, NSF, and the Foundation for Health Environment Research.

**Sara J. Czaja** Ph.D., is the Gladys and Roland Harriman Professor of Medicine and Director of the Center on Aging and Behavioral Research at Weill Cornell Medicine. She is also the Director of the multi-site Center for Research and Education on Aging and Technology Enhancement (CREATE), which focuses on the interface between older adults and technology systems. CREATE is funded by the National Institute on Aging. She is also the Co-director of the Enhancing Neurocognitive Health, Abilities, Networks, and Community Engagement (ENHANCE) Center, funded by the National Institute on Disability, Independent Living, and Rehabilitation Research (NIDILRR), which focuses on technology support for older adults living with a cognitive impairment. She received B.S., M.S., and Ph.D. degrees from the University of Buffalo in New York and is an internationally recognized behavioral scientist with a background in human factors engineering, gerontology, and psychology. She was recruited to Weill Cornell Medicine from the University of Miami Miller School of Medicine where she was Director of the Center on Aging. She is a Fellow of the American Psychological Association (APA), the Human Factors and Ergonomics Society (HFES) and the Gerontological Society of America (GSA). Dr. Czaja is the recipient of the 2015 M. Powell Lawton Distinguished Contribution Award for Applied Gerontology of APA, the 2013 Social Impact Award for the Association of Computing Machinery, the 2013 Jack A. Kraft Award for Innovation from HFES, the Franklin V. Taylor Award of APA, and the 2020 M. Powell Lawton Award of GSA. In addition, CREATE was the recipient of the first American Psychological Association Prize for Interdisciplinary Team Research.

# Introduction 1

## 1.1 The Promise of AgeTech

We are living in an era of rapid technological innovation. This innovation presents novel opportunities for technology to foster and enhance the independence, productivity, health, safety, social connectivity, and quality of life of older people. This book series defines *AgeTech* specifically as "the use of advanced technologies such as information and communication technologies (ICTs), robotics, mobile technologies, artificial intelligence (AI), ambient systems, and pervasive computing to drive technology-based innovations to benefit older adults" (Sixsmith et al., 2020). Both established and emerging technologies can play significant roles in achieving these objectives. However, as we will discuss throughout this book, for these technologies to reach their fullest potential they must be designed with a user-centered design approach that considers the needs, abilities, and preferences of diverse populations of older people. They should be evaluated for usability, and, eventually, be rigorously assessed to determine if they are achieving their intended outcomes and if their use is associated with any unintended consequences.

Charness (2020) elaborates in his *Prevent*, *Rehabilitate*, *Augment*, and *Substitute* (PRAS) framework that there are four primary roles technology-based interventions can play in aiding older people. First, technology can facilitate activities that prevent adverse outcomes (e.g., illness, injury, cognitive decline) from occurring in the first place. Second, if prevention is unsuccessful or impossible, technology can assist in the rehabilitation and restoration of functions impacted by age-related changes or disease processes. Third, if rehabilitation is unsuccessful or not feasible, technology can enhance an individual's capacity to perform essential tasks by leveraging and supporting the functionality they still possess. Finally, if augmentation is not viable, technology can substitute for a completely lost function, performing tasks on behalf of older adults that they can no longer perform themselves. These four strategies ultimately contribute to successful aging, defined as



W. R. Boot et al., *Extended Reality Solutions to Support Older Adults*, Synthesis Lectures on Technology and Health, https://doi.org/10.1007/978-3-031-69220-8_1

the ability of individuals to consistently pursue and achieve goals that are important to them throughout their lifespan. The appropriateness of the different strategies (prevent, rehabilitate, augment, substitute) depends on both an individual's level of cognitive and functional impairment and their preference for support. PRAS serves as a useful general framework for understanding the potential of any technology to contribute to successful longevity.

## 1.2 Introducing Extended Reality (XR)

Due to hardware and software advances, we are on the precipice of a potentially significant shift in how technology can shape our interactions with the world around us, and how technology can support older adults, including older adults experiencing cognitive impairment. This book discusses the potential of extended reality (XR). XR serves as an umbrella term encompassing both augmented reality (AR) and virtual reality (VR). Despite the existence of these technologies in various forms for several decades, we have reached a point in time where they have attained a level of sophistication and relative affordability that enables them to offer practical solutions to the needs of older people.

AR refers to the integration of digital information with the physical world in real time. It layers digital enhancements, such as images, sounds, and data, onto our existing surroundings, thereby augmenting our perception of reality (Fig. 1.1). AR can be implemented in several ways, including through mobile smartphone applications that utilize the phone's camera to capture and display the real-world view on screen while overlaying virtual elements onto this image, or through specialized hardware like headsets and glasses that combine real and virtual world elements. Imagine, for example, an older adult with spatial navigation problems due to a stroke. A digital path overlaid onto the real world presented through a pair of AR glasses might guide them home in real time, without the need to interpret complex maps or navigation instructions.

On the other hand, VR can provide a fully immersive digital experience, disconnecting the user from the physical world and placing them within a simulated environment. This is often accomplished with a headset containing 3D near-eye displays and body tracking capabilities. The immersive and compelling nature of VR could have many benefits for older people. Imagine, for example, an older adult going through a stressful life transition, such as the onset of caregiving responsibilities for a loved one diagnosed with dementia. To help manage their stress and support their wellbeing, VR might guide them through breathing exercises within a calming, relaxing, and immersive natural environment. Imagine, as well, how VR might connect an adult child with a parent with dementia in an assisted living facility, providing them the opportunity to engage in meaningful cognitive and social activities, such as museum tours, together despite residing in different cities (Chap. 3).

**Fig. 1.1** Prototype augmented reality (AR) glasses demonstrating AR navigation capabilities (Shutterstock)

## 1.3 Why XR for AgeTech?

While these technologies present exciting and novel possibilities, it is essential to ask, "Why choose XR?" specifically within the context of AgeTech. We make the argument that, for several reasons, XR is uniquely suited to meet the needs of older people with and without cognitive impairments.

AR solutions, for example, align closely with the classic Human Factors Engineering principle of "providing knowledge in the world," which posits that individuals are more successful in completing complex and demanding tasks when information is supplied directly in the task environment, rather than requiring them to learn and recall information from memory. AR takes abstract or invisible information and makes it visible, contextual, and interactive, thereby reducing cognitive load and enabling more intuitive and successful task learning and execution. Adherence to this principle may be particularly beneficial for older adults experiencing memory problems or other cognitive deficits. Furthermore, potential uses of AR align well with the Environmental Support Hypothesis of Morrow and Rogers (2008). This framework posits that performance differences between younger and older people can be minimized or eliminated when appropriate support, external to the task itself, is provided in the environment. Through AR technology, this support can take

**Fig. 1.2** An older adult wearing an immersive virtual reality (VR) headset (Shutterstock)

the form of providing cues, information, and guidance that can facilitate the performance of important everyday tasks.

VR also boasts a unique constellation of properties that can make it a more effective tool for supporting older adults in their everyday lives. VR is particularly suitable for enhancing skill learning and relearning (e.g., after cognitive or functional impairment) among older people. For example, VR has the potential to effectively deliver cognitive and physical rehabilitation following a stroke. VR allows for experiential, active learning in a safe, simulated environment, consistent with methods that foster both the acquisition and retention of skill. The opportunity for repeated practice in a controlled environment offered by VR can also reduce anxiety and enhance learning outcomes. VR can provide real-time, personalized feedback, and adaptive training, accommodating individual differences in learning rate, which can be substantial among the older adult population. Finally, the realistic and immersive simulations that VR provides can enhance the transfer of training, facilitating the application of learned skills to similar real-world situations. The chapters of this book expand upon how XR is particularly well suited as a form of AgeTech. Although AR and VR can take many forms (e.g., implemented via smartphone apps or projection systems), for the most part, our focus will be on applications involving XR headsets (Fig. 1.2).

## 1.4 Overview of the Book

To effectively engage a diverse range of stakeholders from various backgrounds and disciplines that have interests in supporting older adults, we have deliberately steered clear of an excessively academic style in our book. Our intention is to reach multiple audiences, including researchers, technology developers, clinicians, and service providers, inspiring them to contemplate the ways in which emerging XR technologies can address the needs of both cognitively impaired and unimpaired older individuals. Recognizing the heterogeneous nature of our target readership, we have made a conscious effort to avoid technical jargon and provide only key references, though all statements made are backed by empirical support. By adopting a more accessible, readable, and inclusive approach, we aim to ensure that our message resonates with all those who play or wish to play a role in enhancing the wellbeing and quality of life of older adults through technology.

The initial chapters (this chapter and 2) provide background information on the characteristics and potential of XR. They also delve into the characteristics, needs, and preferences of older individuals, emphasizing the significance of a user-centered design process that incorporates older adults at every stage of development. Chapters 3–8 focus on specific "challenge areas" where XR solutions are needed, such as staying connected and promoting healthy lifestyles and wellness. Challenge areas focus on daily activities and tasks that are critical to independence and wellbeing but can become more difficult with age. Most chapters center around a realistic persona, enabling readers to better understand the challenges faced by older adults, the diverse living contexts they inhabit, and the wide range of needs and abilities of older adult populations. After introducing the challenge area and persona, along with their respective needs, we explore how one or more XR solutions can potentially address those challenges. Additionally, we provide a summary of existing research conducted on the potential of XR within each context. Concluding each chapter, we outline crucial next steps for the development and evaluation of the proposed XR solutions, underscoring the importance of ongoing progress in this field. After these challenge area chapters, we provide concrete considerations for the development and design of XR applications.

## 1.5 XR and the Future

It is important to acknowledge that our exploration of the potential of XR to enhance the lives of older adults is taking place in the year 2024. We recognize that the technology landscape is ever-evolving, and we anticipate rapid advancements in XR hardware and software in the coming years. For example, during the writing of the first draft of this book, the Meta Quest 3 was released. Compared to its predecessor, the Quest 2, the Quest 3 features a significantly more powerful processor, a higher display resolution, clearer lenses, and the addition of a color camera, allowing the system to provide both compelling VR and AR experiences. More notably, between the first draft and the final draft of this book, Apple released the Apple Vision Pro (to mixed reviews), marking a major advancement in commercially available AR. Marketed as a "spatial computer," the Vision Pro is a headset that allows users to interact with apps that appear to float in the air within the user's environment. These advancements will undoubtedly provide new and exciting opportunities for VR and AR to enhance the safety, health, wellbeing, quality of life, and social connectivity of older individuals, including those with and without cognitive impairments. However, amidst this changing technological landscape, we wish to emphasize in this book the important needs of older adults that these technologies have the potential to meet, and the enduring significance of principles that will continue to be relevant as technology changes (for example, inclusive and iterative user-centered design, followed by rigorous testing for intervention efficacy). By keeping these principles at the forefront, we can help ensure that insights gained from our discussion are relevant now and will remain so well into the future. It is critical that, as these technologies advance, developers recognize older adults, including those with cognitive impairments, as users who could greatly benefit from them. This is as long as these technologies are developed and designed with their needs, abilities, and preferences in mind.

## References

Charness, N. (2020). A framework for choosing technology interventions to promote successful longevity: Prevent, rehabilitate, augment, substitute (PRAS). *Gerontology, 66*(2), 169–175.

Morrow, D. G., & Rogers, W. A. (2008). Environmental support: An integrative framework. *Human Factors, 50*(4), 589–613.

Sixsmith, A., Sixsmith, J., Fang, M. L., & Horst, B. (2020). Agetech, cognitive health, and dementia. *Synthesis Lectures on Assistive, Rehabilitative, and Health-Preserving Technologies, 9*(2), i–166.

# Designing Technology for Older Adults

# 2

## 2.1 Introduction

The first rule of good design is "know thy user." Since this book centers on developing extended reality (XR) solutions for older adults, it is essential to begin with a robust understanding of the needs, preferences, and abilities of the growing and diverse older adult population. After discussing the characteristics of older people, the challenges they face, and the age-related "digital divide," we outline design principles and methods to ensure that XR technology solutions are useful to and usable by older adults. The chapter provides a broad context for each of the subsequent chapters. We must start with an important caveat: although older adults differ from younger adults on average, there is substantial variability among older people. Older adults are not a homogenous group, and while it is important to acknowledge that age-related differences should impact design considerations, it is also essential to recognize the diversity of the people for whom we are designing.

## 2.2 Population Aging

Many countries in the world are experiencing the phenomenon of population aging. The World Health Organization (WHO) projects that the population of older adults (defined by WHO as individuals 60 years of age or older) will increase to 2.1 billion by 2050, effectively doubling the number of people in this age group worldwide (World Health Organization, 2022). Moreover, from 2020 to 2050, the population of individuals 80 years of age or older will triple. While increases in life expectancy should be celebrated, age-related health and cognitive challenges, along with demographic shifts, are anticipated to place enormous strain on social, healthcare, and economic systems. For example, 80% of



W. R. Boot et al., *Extended Reality Solutions to Support Older Adults*, Synthesis Lectures on Technology and Health, https://doi.org/10.1007/978-3-031-69220-8_2

adults aged 65 and older have at least one chronic condition and 68% have two or more (National Council on Aging, 2021). By 2050, the worldwide number of people living with dementia will rise from about 57 million in 2019 to approximately 153 million (Schwarzinger & Dufouil, 2022). As the population ages and the demand for care and caregiving resources rises, there will be relatively fewer younger people to provide this support due to declining birth rates and changes in family structures (e.g., geographical dispersion). The use of technology can help bridge the gap between societal needs and available resources.

Apart from addressing societal needs, it is beneficial for product developers, designers, and companies to recognize these shifting demographics and to understand changes in their consumer base. As the world ages, so does the population of people using and purchasing products. From a business standpoint, the increasing number of older adults presents a significant market shift that can't be ignored. To stay competitive and relevant, designers and marketers need to tailor their products and strategies to meet the needs and preferences of this demographic. Businesses that acknowledge and embrace this change will hold a competitive edge in the evolving marketplace.

Designing technology for older adults is therefore a crucial goal, and doing so effectively requires the cultivation of knowledge about how people change as they age. In the following sections, we briefly review some of these changes and consider the wide variety of differences between younger and older people. These changes and differences inform not just design considerations, but activities technologies should support to foster independence among aging adults.

## 2.3 Age-Related Changes in Function

As we age, several changes can occur that impact function and the ability to use various technologies. We provide a brief overview of these important changes in this section. Although we will be discussing age-related changes, as mentioned previously, it is important to point out that there is tremendous variability among older people at the rate in which and how they age.

*Sensation and Perception.* Sensation refers to how our body gathers signals from the world around us through our senses, like touch, sight, and hearing. Perception, on the other hand, is how our brain interprets these signals, turning them into images, sounds, and other things we recognize. Age affects both processes. Visual impairments increase with age, often impacting visual acuity, the ability to see detail and contrast between visual images (e.g., text against the background of a computer screen). Additionally, visual declines make it more challenging to extract images in the periphery, leading to a reduced field of view. Beyond vision, hearing loss is common with age, affecting high-frequency sounds particularly, and men more than women. Due to age-related hearing loss, understanding speech in noisy environments can be especially challenging. Declines in sensitivity to touch are also common, with older adults being less sensitive to vibration, especially high-frequency vibration. Although less relevant to design, age-related changes in taste and smell can also occur. In a very real sense, older adults see, hear, and feel the world differently compared to younger adults, and for them to be able to extract the same amount of information as younger people sensory signals may need to be stronger and they may need longer to process the information.

*Cognition.* After sensing and perceiving information in the environment, cognition refers to the process by which our brain stores, retrieves, transforms, understands, reasons, and ultimately acts upon this information. Cognitive aging is a well-documented phenomenon; many cognitive abilities change as a natural part of the aging process. Much of the declines observed in cognition can be attributed to age-related changes in processing speed, the speed with which individuals can perform the elemental cognitive steps that underlie complex cognitive processing. Cognitive changes include age-related changes in attention, the ability to prioritize the processing of information relevant to an ongoing task, and changes in inhibitory control, the ability to ignore distracting information. Cognitive changes also include declines in working memory capacity, which refers to the ability to temporarily store and manipulate information in the mind, which can decline substantially with age. Executive control, which involves maintaining and updating goals, planning and sequencing actions, problem-solving, and inhibiting automatic responses, decreases significantly. In contrast, semantic knowledge, or knowledge about the world, tends to either stay the same or increase later in life. Experience and learned strategies can help offset cognitive ability declines to some degree. Good design for older adults should account for age-related changes in the way individuals process, store, and retrieve information. Technology design that places large memory demands on the user or requires rapid cognitive processing will disadvantage older adults especially.

*Cognitive Impairment, Dementia, and Alzheimer's Disease.* Some individuals will experience age-related injury or disease processes that impact cognitive abilities to a much greater extent compared to typical cognitive aging. Dementia is an umbrella term encompassing conditions that lead to a substantial decrease in cognitive function that interferes with daily life. Types of dementia include Alzheimer's disease, vascular dementia, and Lewy body dementia, which are neurodegenerative conditions that lead to progressive and severe cognitive decline. Frequently, these more serious declines are preceded by a state known as mild cognitive impairment (MCI), a condition in which an individual experiences cognitive declines that are more pronounced than typical cognitive aging, but not severe enough to be classified as dementia. Many individuals with MCI subsequently experience further decline. Age-related injuries such as stroke can also impact cognitive abilities depending on the severity and location of the brain injury. Furthermore, older individuals are prone to traumatic brain injury (TBI) due to events such as falls or car accidents. TBI can induce both short-term and long-term cognitive deficits. Cognitive ability changes due to disease or injury, more so than typical cognitive aging, threaten independence, wellbeing, and quality of life, and technology has tremendous potential to help remediate or compensate for these changes.

*Movement Control, Strength, and Range of Motion.* As people age, movement control—the coordination of muscles for action—typically deteriorates, resulting in slower and less precise movements in older adults compared to their younger counterparts. Age-related changes in muscle and connective tissue can limit the range of various motions, and declines in muscle mass can decrease strength. Factors such as reduced hand-grip strength, diminished grip endurance, and age-related tremors can significantly influence an older adult's interaction with everyday technologies. Disease processes, such as arthritis, can exacerbate the impact of these changes. However, declines in strength and movement control will vary substantially among older adults. Regular exercise can help mitigate many of these adverse effects of aging. As will be discussed in Chap. 4, technology solutions hold tremendous potential for encouraging and facilitating physical exercise. Regardless, technology designers need to account for these changes to ensure successful system interactions across all age groups. Considerations could be as simple as ensuring that device buttons are large and well-spaced or incorporating voice commands to largely bypass age-related changes in movement control.

## 2.4 The Age-Related Digital Divide

Despite recent increases in the adoption of technologies by older adults, a significant age-related "digital divide" persists. This divide refers to the differences between younger and older people in their ownership, use, and proficiency with a variety of technologies. For instance, while the use of the Internet among younger adults is nearly universal in the United States, twelve percent of older adults are still not online as of 2023 (Pew Research Center, n.d.). Nearly a quarter (24%) of older adults did not own a smartphone, compared to only 3% of individuals aged 18–29. Even among older adult smartphone owners, their proficiency in using smartphones can be substantially lower compared to younger adults (Roque & Boot, 2018). Similar trends in ownership, use, and proficiency are observed for a variety of existing and emerging technologies, including social media platforms, wearable devices, smart home devices, and digital voice assistants (e.g., Alexa). The consequence of this divide is that many older adults are not afforded the advantages provided by technologies that could help support their health, wellbeing, and quality of life. Within this older adult cohort, individuals experiencing disability, who are older, who receive a lower income, or who live in rural areas, are especially disadvantaged in terms of technology access and use. Technology proficiency differences between younger and older people are even more pronounced among older adults experiencing even mild cognitive impairment (Schmidt & Wahl, 2019).

The age-related digital divide is influenced by several factors, particularly attitudinal, cognitive, health, and perceptual, and design barriers. Attitudinal barriers arise from older adults' beliefs about their ability to use and their interest in using digital technologies. Differences in attitudes between younger and older adults may not relate to age-related changes per se, but to the fact that different generations grew up using different technologies. Although these attitudes are modifiable, they may require targeted interventions that emphasize the benefits of technology use. Training and instructional support can also facilitate successful technology interactions and influence older adults' beliefs about their ability to use technology. These approaches are consistent with models of technology use and adoption that emphasize perceived usefulness and ease of use as primary factors that drive whether a person will consider using a new piece of technology. Finally, addressing privacy and security worries is crucial for technology acceptance. Developers must establish strong security protocols and clearly inform users about these safeguards to enhance trust and promote usage.

As previously mentioned, cognitive and perceptual barriers stem from natural changes in cognitive and perceptual abilities as people age, making the use of various technologies more challenging. The cost of new learning is often higher for older adults, and as a result they typically require more time and effort to understand and operate new technologies compared to younger individuals. This can disincentivize technology adoption. Finally, design barriers arise when technology developers fail to account for the specific needs, preferences, and abilities of older people. This lack of consideration often results in perceptions among older adults that technologies are less useful or harder to use. A contributing factor to this issue may be the disparity in age between those creating technology and the older adult users. Younger designers may find it challenging to empathize with and design for the experiences and needs of aging adults due to an inherent egocentric bias most people experience when it comes to understanding other individuals' knowledge and abilities.

## 2.5 Design Frameworks

The process of designing technology for older adults, to be successful, needs to be systematic, thoughtful, and informed by potential users at all stages of the design process. Key models for designing technology include the Nielsen Norman Group's *Design Thinking 101* framework, which promotes a user-centered, iterative, and empirical approach (Nielsen Norman Group, n.d.). This process begins with understanding user needs and experiences, refining problems, brainstorming solutions, and testing prototypes with target users in advance of implementation. The process can be repeated and revisited at any stage to ensure the technology solution addresses real needs and is usable by the target group, such as older adults. It encourages generating numerous solutions before selecting one which will be developed further, which fosters innovation.

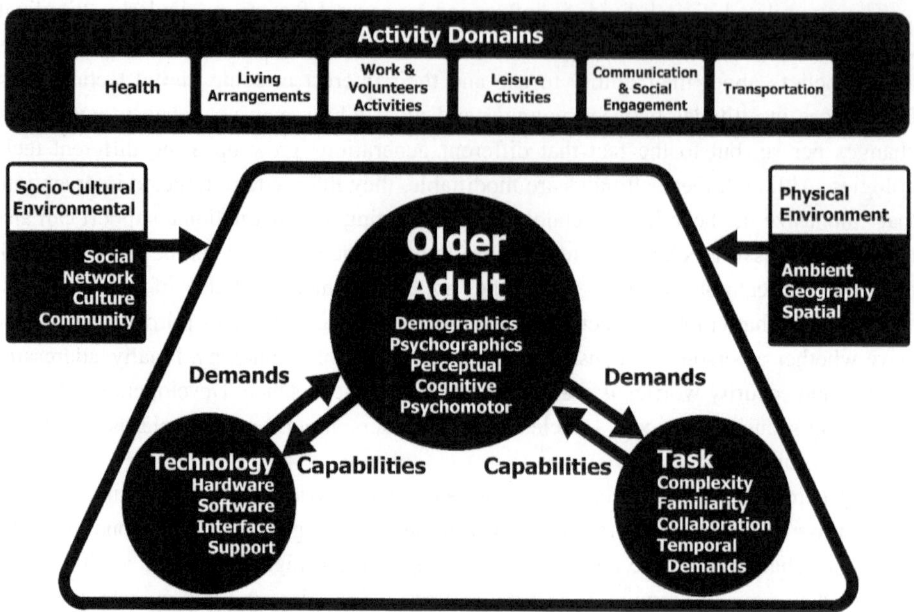


**Fig. 2.1** The CREATE model of design for older adults

Person-environment fit-based frameworks, like the one developed by the Center for Research and Education on Aging and Technology Enhancement (CREATE), are also crucial (Fig. 2.1). They focus on aligning the demands the technology places on the user with the user's capabilities. The CREATE model situates the older adult at the center of the design process, and recognizes their cognitive, perceptual, and motor abilities, all of which impact their ability to use technology. It also considers their technological experience and proficiency. Additionally, the model recognizes contextual factors such as available support and environmental elements (e.g., lighting, noise). Successful technology solutions for older adults arise from ensuring a match between system demands and user capabilities, thereby preventing frustrating user experiences, and developing and implementing a supportive context to facilitate technology learning and use.

## 2.6 Design Considerations and Methods

We recommend implementing a variety of methods during the design process to achieve a "fit" between technology demands and users' capabilities. When using these methods, it is critically important to include representative samples of aging adults that are diverse on attributes important to the use of the technology of interest. Surveys, focus groups, and interviews often provide key insights into the everyday challenges that older adults experience, especially when the participants include a wide range of older adults, such as those with and without cognitive impairment. For example, in designing XR solutions to support physical exercise, these methods can help uncover the challenges older adults currently experience in getting the recommended amount of physical activity. This "needs assessment" phase ensures that designers address real problems that older adults face, rather than problems the designers presume exist. It can provide a deep understanding of the problem and older adults' preferences for solutions to guide technology development.

Once the needs of the users are established and their preferences for technological support are understood, the brainstorming process for technology solutions can begin. At this stage, older adults might be shown various prototypes of the system (sometimes as simple as a sketch on paper) to elicit initial feedback that can be incorporated into the system design process. For example, they could be shown mockups of various XR exergames aimed at improving physical fitness. Design should be iterative, with prototypes modified based on initial feedback from representative users.

The design of the technology itself should then consider the various age-related changes discussed previously in this chapter. For example, in the development of an XR solution involving a handheld controllers, the designer should contemplate whether age-related changes in motor control and disease processes such as arthritis might make the controllers hard to use for some older adults, and how alternative designs might help mitigate this issue. If controller vibration is used to provide feedback to the user as to whether this vibration cue is strong enough and within the right range of frequencies to promote detection given known age-related changes. Designers should think carefully about how declines in sensory and cognitive processing might be accommodated in their design. XR app menus should feature large and high-contrast fonts, and menu options should be presented in a way that does not place high demands on working memory. For more comprehensive discussion of guidelines on designing for older adults, we direct readers to *Designing for Older Adults: Principles and Creative Human Factors Approaches, Third Edition* (Czaja et al., 2019). Designing while being mindful of age-related changes will help ensure the ultimate success of XR solutions.

As the system development progresses, usability testing can begin, which involves evaluating older adults' interactions with the system. This testing aims to assess several key measures to ensure a positive user experience. “Learnability” measures how easily a new user can understand and use a product or system. “Efficiency” pertains to the speed with which tasks can be performed once the user is familiar with the interface. “Memorability” assesses how easily a user can reestablish proficiency after a period of non-use. Other important considerations are the frequency, severity, and recoverability of errors made by users, as well as user satisfaction, which gauges how happy users are with the overall system or product. Any issues uncovered during this process can be addressed, and the system can undergo redesign, followed by another round of usability checks.

To increase the efficiency of usability testing, several methods can be deployed in advance of having older adults interact with the technology system. For example, a heuristic analysis compares aspects of the technology against established rules of good design, referred to as usability heuristics. One such heuristic is visibility of system status, which assesses how well the system keeps users informed about what's happening through appropriate feedback. Another is match between system and the real world, which measures the extent to which the system uses words, phrases, and concepts familiar to the user. By checking a design's adherence to usability heuristics, obvious design flaws can be uncovered and corrected without the need for human participants. Similarly, the cognitive walkthrough methodology involves expert evaluators performing and analyzing various tasks that the user can perform with the system, step by step, while considering the knowledge and abilities of the user, and whether there is sufficient information presented by the system to guide the user to the next correct action to accomplish their goal. If there is not sufficient information to guide the user, redesign should be considered. These initial systematic approaches can improve system design, allowing usability testing to focus on uncovering more subtle system flaws.

A final critical step involves confirming the efficacy of the developed technology-based solution. In other words, does the implementation of this solution in the context of an older adult's life bring about benefits or unintended consequences? For instance, a randomized controlled trial might be conducted in which some older adults receive the technology while others do not. Outcome measures of interest, such as physical fitness, cognitive ability, and loneliness, might be assessed before and after the intervention period to explore whether those receiving the technology improved differentially. Well-designed efficacy trials can help determine not only whether the intervention works, but why it works (uncovering mechanism of action), as well as who benefits most from the technology.

## 2.7 Summary

XR solutions rely on sophisticated and complex technology. As of now, many older adults may be unfamiliar with this technology, and some may be reluctant to adopt and use it. A variety of strategies, however, are likely to facilitate adoption and use. First, if XR-based solutions are to be beneficial to older adults, their designs must consider the needs, preferences, and abilities of diverse groups of older people. Second, usability should be rigorously evaluated using appropriate methodologies. Third, efficacy testing is important to understand the potential benefits of the developed technology solution. Additional barriers to use and adoption may be overcome through training and the provision of appropriate instructional support. Adherence to technology development and design frameworks, coupled with a user-centered, iterative design process, and followed by efficacy testing, is crucial to ensuring that XR solutions are both useful to and usable by diverse groups of older people.

## References

Czaja S. J., Boot W. R., Charness N., & Rogers W. A. (2019). *Designing for older adults: Principles and creative human factors approaches* (3rd ed.). Taylor and Francis.

National Council on Aging (2021). *The top 10 most common chronic conditions in older adults.* Retrieved from 11 July 2023: https://www.ncoa.org/article/the-top-10-most-common-chronic-conditions-in-older-adults

Nielsen Norman Group. (n.d.). *Design thinking 101.* Retrieved from 01 July 2023: https://www.nngroup.com/articles/design-thinking/

Pew Research Center. (n.d.). *Internet/broadband fact sheet.* Retrieved from 16 March 2024: https://www.pewresearch.org/internet/fact-sheet/internet-broadband/

Roque, N. A., & Boot, W. R. (2018). A new tool for assessing mobile device proficiency in older adults: The mobile device proficiency questionnaire. *Journal of Applied Gerontology, 37*(2), 131–156.

Schwarzinger, M., & Dufouil, C. (2022). Forecasting the prevalence of dementia. *The Lancet Public Health, 7*(2), e94–e95.

Schmidt, L. I., & Wahl, H. W. (2019). Predictors of performance in everyday technology tasks in older adults with and without mild cognitive impairment. *The Gerontologist, 59*(1), 90–100.

World Health Organization. (2022). *Ageing and health.* Retrieved 15 June 2023 from: https://www.who.int/news-room/fact-sheets/detail/ageing-and-health

# Staying Connected 3

## 3.1 The Challenge

In 2023, U.S. Surgeon General Dr. Vivek Murthy warned of an epidemic of social isolation and loneliness and noted that many older people face negative consequences because of this public health threat (U.S. Public Health Service, 2023). Although closely related, social isolation refers to a lack of social connections, which puts individuals at greater risk for loneliness, the subjective feeling of being alone. Both isolation and loneliness are associated with adverse outcomes. Social isolation has been linked to numerous negative consequences, such as lower quality of life, reduced life satisfaction, and poorer mental and physical health. Social isolation is also associated with increased risk of mortality. Analyses have indicated that prolonged isolation and loneliness can be as damaging to one's health as smoking (Holt-Lunstad et al., 2010). Fortunately, technology-based solutions have the potential to intervene and reduce or prevent these negative effects.

Among older adults, isolation and loneliness can arise from a variety of factors, including reduced mobility, health issues, changing financial and employment status, and the death of a partner and friends. Older people experiencing cognitive impairment face an even greater risk for social isolation and loneliness. Cognitive challenges can make maintaining relationships difficult, causing individuals to socially withdraw. Withdrawal can lead to greater isolation and loneliness, further exacerbating cognitive decline. In sum, a variety of age-related changes and challenges can lead to a smaller social network size and increased feelings of loneliness, especially among those who live alone. In the United States, about 27% of older adults and 43% of women over the age of 75 live alone (Administration on Aging, 2022).

These facts highlight the urgent need for strategies to combat social isolation in older adults. The use of information and communication technologies to promote social connections, such as the Internet or email, is a promising solution. By bridging physical



W. R. Boot et al., *Extended Reality Solutions to Support Older Adults*, Synthesis Lectures on Technology and Health, https://doi.org/10.1007/978-3-031-69220-8_3

distances and overcoming mobility challenges, these tools can foster social inclusion and connectivity, helping to mitigate the damaging effects of isolation and staving off loneliness. The National Institute on Aging, in their *Understanding Loneliness and Social Isolation* guidelines, recommends a variety of technology-mediated communication methods to help older adults stay connected, including video chat software (National Institute on Aging, 2020). Emerging technologies offer even newer methods of communication, including extended reality (XR) communication solutions. These solutions have the potential to further expand the ability of socially isolated older adults to engage in frequent and meaningful social interactions. XR has tremendous potential to support social presence, the feeling that someone at a distance is inhabiting the same space.

## 3.2 What's in This Chapter?

This chapter reviews some of the challenges related to maintaining social connections as we age. The potential of augmented reality (AR) and virtual reality (VR) to address these issues is discussed, as well as a brief review of the literature highlighting the promise and pitfalls of these approaches. These issues are presented in the context of the persona Joe, a man experiencing isolation and loneliness as a result of several age-related life changes (Fig. 3.1).

**Fig. 3.1** Joe, a widower who lives alone with chronic health conditions, has been lonely since his retirement (Shutterstock)

## 3.3 Persona and Scenario

**Persona**: Joe is a 72-year-old widowed man living alone in a home in Woodstock, a suburb of Chicago. Throughout his career as a librarian, his job provided him with a sense of purpose and a social circle. While some people flourish post retirement, Joe's life has grown increasingly solitary since leaving the workforce at the age of 68. He enjoyed his job and would have liked to have worked longer, but opted for retirement because of mounting health challenges, including a recent diagnosis of early-stage heart failure. Despite his declining health, he is still able to manage most daily activities within his home, but he has withdrawn from activities outside of the home. When he does leave home, it is typically for doctors' appointments. Joe often relies on delivery services for groceries and prescriptions and can go weeks without talking to another person face-to-face. Joe feels himself becoming increasingly disconnected from the world and bored since his retirement and the onset of his health problems. The one bright spot in his social life is his weekly call with his only child, Melissa. Unfortunately, Melissa works in San Francisco, and due to her busy career and family life, she can only visit once or twice a year. When she does make the trip, Joe cherishes their time together. When his health was better, they would visit the art and history museums in Chicago. Joe has always been a social person, but like many older adults, changing life circumstances have reduced the size of his social network and the frequency of his social contacts. In Joe's case, this has led to profound loneliness some days.

**Scenario**: Retirement and the loss of his spouse have contributed to Joe's social isolation. Unfortunately, his isolation is exacerbated by chronic health conditions that have resulted in mobility challenges. These challenges present major barriers to visiting others, making new friends, and attending social events. He is grateful for Melissa's phone calls and wishes he could see her more often, but he understands her obligations. He misses the trips they used to take to explore new museum exhibits in Chicago and mentions this to her during one of their weekly calls. Joe desires richer, more frequent, and more meaningful social interactions in his life.

**Potential Solutions**: Melissa gifted her father with a VR headset for Christmas during her most recent visit to Woodstock. After setting up the device, she downloaded a program that would allow them to take virtual museum and sightseeing tours together. She also helped Joe complete VR tutorials to learn how to use the system and software. Following her return to San Francisco, Joe and Melissa set up a time to meet at the virtual art museum. Melissa donned her headset in San Francisco and Joe did the same in Woodstock. The app then co-located them into the same virtual art gallery, and they took the tour together. They could see one another in the virtual space, depicted as avatars, and talk to each other. Each work of art was accompanied by a virtual plaque providing descriptions of the pieces and their respective artists. They navigated the gallery together taking turns reading the plaques aloud. They discussed what they liked and disliked about the paintings and sculptures and their interpretation of what the artist was trying to convey. Before they knew it, an hour had passed.

Next time, they took a virtual tour of Paris, which included a view from the top of the Eiffel Tower. Joe looks forward to these "visits" because it feels as if he and Melissa are in the same space, sharing the same experience. He also started playing games with his grandchildren in VR. After a while, Joe began to explore other applications that allowed him to engage in social interactions beyond his immediate family. For Joe, VR contributed to an enhanced sense of wellbeing and reduced loneliness, overcoming many of the barriers to meaningful social interactions he previously faced.

## 3.4 XR Solutions for Staying Connected

VR- and AR-based social applications hold promise for helping older adults connect with others. These apps can create opportunities for meaningful social connections and offer enriching interactions that may be difficult for older adults to achieve outside of XR due to a variety of age-related challenges. AR might further enhance social experiences by placing a virtual social partner within an individual's physical surroundings, offering an even more personal and intimate connection (e.g., Miller et al., 2019). By enhancing social interactions and expanding social networks, these technologies have massive potential to combat isolation and loneliness among older adults.

Currently, however, there are few if any rigorous, large-scale intervention studies that have examined whether XR-mediated social interactions can boost social support and reduce feelings of loneliness, especially among older people. However, initial studies support the feasibility and acceptability of this approach. In a small study, we paired older adults from three cities and placed pairs of participants into the same virtual space using the Meta Quest 2 (Kalantari et al., 2023). We designed a unique social VR environment, integrating features intended to stimulate conversation and encourage cooperative problem-solving (Box 3.1). The centerpiece of this experience was a virtual travel module that allowed participants to discuss and then choose an international destination to visit together. Once selected, participants could tour the destination city with their partner, and see and talk to their partner throughout the tour. Of primary interest was whether older adults would find these experiences engaging and whether they would enjoy their interactions with their virtual partner. The experiences of older participants were almost universally positive. They found the modules to be highly engaging, and most older adults reported a willingness to reconnect with their social partner again. For individuals such as Joe, these types of programs might provide not just social support, but meaningful activity engagement in his retirement years. However, as outlined below in the Key Initiatives section, additional research is needed to fully understand the potential benefits and unintended consequences of this approach.

In another recent small-scale study, researchers explored the effectiveness of VR in enhancing the lives of older adults in a senior living community (Afifi et al., 2023). This study included older adults experiencing cognitive impairments, some of whom lived in assisted living, and connected older adults with remote family members via VR. A total of 21 older adults with cognitive impairment engaged in VR sessions spanning three weeks with a family member (in most cases, an adult child). VR experiences included touring remote travel destinations together, using Google Street View in VR to visit destinations from the participants' pasts, and viewing family photos and videos in VR while sitting on the same virtual couch. Results indicated significant improvements in emotional wellbeing and quality of life for the cognitively impaired older adults after VR exposure. Older adults also reported an increase in emotional closeness and relationship satisfaction with their remote family member. Additionally, their family members experienced a decrease in negative emotions, depressive symptoms, and caregiver burden, alongside a notable uptick in their mental health. Though not statistically significant, trends were consistent with more impaired older adults receiving greater benefits. This study provides compelling evidence for the feasibility, acceptability, and preliminary efficacy of VR to promote social connections and enhance relationship quality and quality of life.

What makes a social VR experience successful? Although the researchers did not focus on older adults specifically, recently van Brakel et al. (2023) administered surveys to users of social VR platforms to better understand what aspects of virtual social interactions are associated with higher levels of perceived social support and wellbeing. They focused on the concepts of *social presence* and *self-presence* as predictors. Social presence is a psychological concept referring to the perceived sense of being co-located and interconnected with another person, often in the context of digital environments. Social presence is related to the sense that one is physically present with another person despite geographical distance. Self-presence, also known as embodiment, relates to a user's perception that their avatar's body in the virtual environment is their actual body. The researchers found that higher levels of social presence and self-presence were associated with higher levels of perceived social support offered by VR, and that perceived social support had a positive relationship with wellbeing. Although there are limitations determining causality with self-report survey data, these results provide initial insight into not only the potential influence of social VR on social support and wellbeing, but also the mechanisms through which perceived social support and wellbeing might be enhanced in the context of social VR platforms.

## 3.5 Key Initiative

Of the existing XR applications aimed at supporting older adults discussed in this book, the domain of assisting with social connections is one of the least explored. However, there are good examples available of how to build and test technology-based interventions aimed at increasing social support and reducing loneliness. One example is the Personal Reminder Information and Social Management (PRISM) randomized controlled trial (Czaja et al., 2018). The aim of this study was to enhance social support and reduce loneliness of older adults living alone and at risk for social isolation. Through an iterative, user-centered design approach, the PRISM software system was developed and featured email, Internet, games, and other functions designed to facilitate social connectivity and resource access. Three hundred older adults either received the PRISM system in their home for a year or were assigned to a notebook condition that contained a lot of the same information as PRISM, but in a non-electronic format. After 6 months, the group that used the PRISM system reported feeling less lonely and reported more social support.

A similar design process and large-scale trial is necessary in order to ensure that XR-based solutions are designed to be useful and usable by diverse populations of older adults. Large trials such as this can also help identify why the intervention works and for whom it works. When it comes to technology to support social connectivity, there is always a concern that rather than supplement and support an individual's existing social connections, technology-based solutions have the potential to replace these connections. This is an unintended consequence that is important to monitor for during future studies. Should this occur, it is possible that social isolation might be exacerbated rather than reduced. XR solutions aimed at helping older adults stay connected need to give careful consideration to the balance between virtual and real-world interactions.

Before large-scale efficacy trials are initiated, smaller needs assessment, usability, feasibility, and acceptability studies are required. Chapter 2 outlines how a user-centered design process, involving diverse stakeholders and potential system users, is critical for the successful deployment of technology-based interventions. This is very much an ongoing effort in the domain of XR to support social engagement. For example, Mikhailova et al. (2024) recently conducted a qualitative interview study to explore older adults' general attitudes and preferences for AR-facilitated communications, using storyboard illustrations to facilitate discussion. Through interviews, Flynn et al. (2024) explored the experiences and viewpoints of individuals with dementia and their caregivers regarding social engagement in general, along with the possible impact of VR in fostering or preserving such connections. Oppert et al. (2023), in a pilot study, assessed the acceptability of a VR social program among a small sample of older adults experiencing loneliness. This is a fast-moving field of research, and these early studies that address needs, attitudes, and preferences of diverse older adults in diverse contexts are crucial for the development of interventions that can then be tested for efficacy in larger scale efficacy trials later.

**Box 3.1 Testing a Social Engagement VR Application**

Kalantari and colleagues developed a novel VR program to offer social opportunities for older adults (Fig. 3.2). The program has four modules. **Training Module**: This is an orientation module where participants familiarize themselves with the VR environment and controls, including avatar creation, movement, and object manipulation, with a moderator available for assistance. **Introductions Module**: This social space module enables participants to meet others, share life and travel experiences, and collaboratively choose a virtual travel destination, guided by a moderator. **Travel Module**: Participants virtually explore the selected location through 360-degree videos of landmarks, engaging in discussion about their experiences while their avatars remain visible to each other. **Productive Engagement Module**: In this module, participants collaboratively perform memory and creativity tasks related to their virtual travel, identifying and arranging photos into a collage, while promoting cognitive engagement and problem-solving.

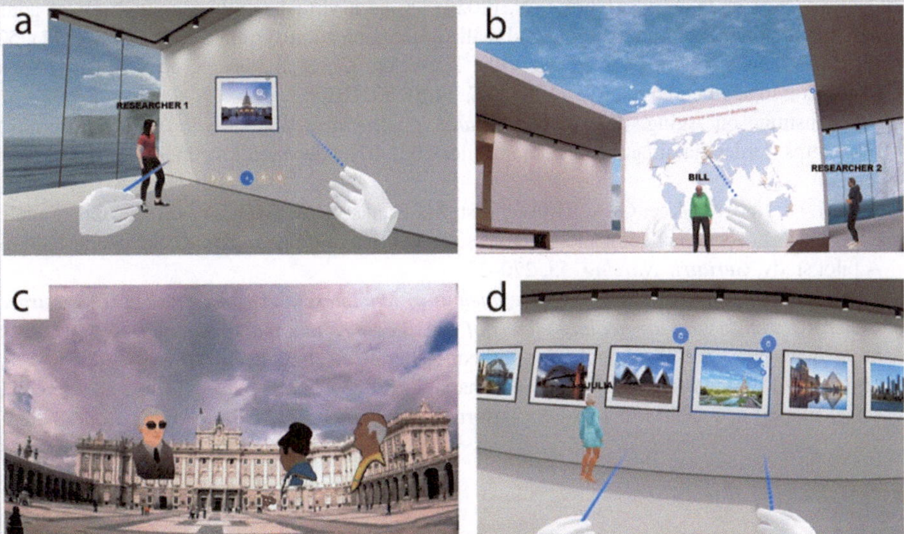


**Fig. 3.2** Social VR experiences that participants engaged in, including the **a** training module, **b** introduction module, **c** travel module, and **d** productive engagement module

## References

Administration on Aging. (2022). 2021 profile of older Americans. Retrieved from https://acl.gov/sites/default/files/Profile%20of%20OA/2021%20Profile%20of%20OA/2021ProfileOlderAmericans_508.pdf

Afifi, T., Collins, N., Rand, K., Otmar, C., Mazur, A., Dunbar, N. E., & Logsdon, R. (2023). Using virtual reality to improve the quality of life of older adults with cognitive impairments and their family members who live at a distance. *Health Communication, 38*(9), 1904–1915.

Czaja, S. J., Boot, W. R., Charness, N., Rogers, W. A., & Sharit, J. (2018). Improving social support for older adults through technology: Findings from the PRISM randomized controlled trial. *The Gerontologist, 58*(3), 467–477.

Flynn, A., Brennan, A., Barry, M., Redfern, S., & Casey, D. (2024). Social connectedness and the role of virtual reality: Experiences and perceptions of people living with dementia and their caregivers. *Disability and Rehabilitation: Assistive Technology*, 1–15.

Holt-Lunstad, J., Smith, T. B., & Layton, J. B. (2010). Social relationships and mortality risk: A meta-analytic review. *PLoS Medicine, 7*(7), e1000316.

Kalantari, S., Xu, T. B., Mostafavi, A., Kim, B., Dilanchian, A., Lee, A., & Czaja, S. J. (2023). Using immersive virtual reality to enhance social interaction among older adults: A cross-site investigation. *Innovation in Aging*, 7(4), igad031.

Mikhailova, V., Conde, M., & Döring, N. (2024). "Like a virtual family reunion": Older adults defining requirements for an augmented reality communication system. *Information, 15*(1), 52.

Miller, M. R., Jun, H., Herrera, F., Yu Villa, J., Welch, G., & Bailenson, J. N. (2019). Social interaction in augmented reality. *PLoS ONE, 14*(5), e0216290.

National Institute on Aging. (2020). *Understanding loneliness and social isolation.* Retrieved from https://order.nia.nih.gov/sites/default/files/2021-01/understand-loneliness-and-social-isolation.pdf

Oppert, M. L., Ngo, M., Lee, G. A., Billinghurst, M., Banks, S., & Tolson, L. (2023). Older adults' experiences of social isolation and loneliness: Can virtual touring increase social connectedness? A pilot study. *Geriatric Nursing, 53*, 270–279.

U.S. Public Health Service. (2023). *Our epidemic of loneliness and isolation. The U.S. Surgeon General's advisory on the healing effects of social connection and community.* Retrieved from https://www.hhs.gov/sites/default/files/surgeon-general-social-connection-advisory.pdf

van Brakel, V., Barreda-Ángeles, M., & Hartmann, T. (2023). Feelings of presence and perceived social support in social virtual reality platforms. *Computers in Human Behavior, 139*, 107523.

# 4 Healthy Lifestyles and Wellness

## 4.1 The Challenge

Mental and physical health are crucial facets of successful aging, and older adults can incorporate a variety of activities into their daily routine to foster wellness across their lifespan. This chapter places a spotlight on two health-promoting activities that may particularly benefit from extended reality (XR) technology: the pursuit of physical exercise and the adoption of relaxation techniques such as meditation.

The Centers for Disease Control and Prevention (CDC) in the United States recommends that each week older adults participate in at least 150 min of moderate-intensity physical activity and engage in specific activities to improve balance and muscle strength (Centers for Disease Control and Prevention, n.d.). Unfortunately, physical activity decreases with age, and many older adults do not exercise as much as recommended. This puts them at a greater risk for conditions such as diabetes, heart disease, and cognitive decline, among other health risks. While it is widely recognized that physical exercise is important, maintaining motivation and adherence to exercise programs can be a challenge for many people, especially over the long term. Many individuals who want to increase their physical activity do not know how to start and would benefit from guidance. Environmental barriers, such as poor weather conditions, lack of proximity to indoor and outdoor exercise facilities and recreation areas, neighborhood safety risks, and ageism can also serve as obstacles to engaging in exercise.



W. R. Boot et al., *Extended Reality Solutions to Support Older Adults*, Synthesis Lectures on Technology and Health, https://doi.org/10.1007/978-3-031-69220-8_4

Additionally, as individuals age, they face several stressful life transitions that can impact their health and wellbeing. Retirement, changes in health status, new caregiving responsibilities, and the loss of a spouse or friends are a few examples of these transitions. Stress is a natural response to these changes, but it can also have negative impacts on mental and physical health. Chronic stress has been conceptualized as a force that accelerates the aging process, making it an important target for interventions to promote successful longevity. For example, stress can cause psychological distress and weaken the immune system, which can lead to a higher risk of illness and disease. It can also harm brain function, contributing to cognitive declines (Prenderville et al., 2015). The National Institute on Aging (NIA) recognizes the importance of managing stress to promote healthy aging and makes several recommendations to older people (National Institute on Aging, n.d.). While exercise is an important component of stress reduction, NIA also recommends engaging in relaxation techniques such as mindfulness and breathing exercises. Meditation, for example, has been shown to have several mental and physical health benefits, including reducing anxiety, improving sleep, and lowering blood pressure. However, many people are unfamiliar with these techniques and would likely benefit from instruction and guidance to get started.

Technology, particularly augmented reality (AR) and virtual reality (VR) solutions, holds tremendous potential for promoting physical exercise and wellness strategies like meditation to enhance mental and physical health. These technology solutions can provide a fun and engaging way for individuals to stay motivated and committed to healthy aging practices over the long term.

## 4.2 What's in This Chapter?

This chapter reviews some of the challenges related to maintaining lifestyle habits to boost health and wellbeing. The potential of AR and VR to address these challenges is explored, as well as a brief review of the literature examining the promise and pitfalls of these approaches. These issues are discussed in the context of the persona Mary, an older woman who lives in an urban area and wants to increase her physical activity level but faces several personal and environmental challenges in doing so (Fig. 4.1).

**Fig. 4.1** Mary seeks a solution that allows her to enjoy a more active lifestyle (Shutterstock)

## 4.3 Persona and Scenario

**Persona**: Mary is a 76-year-old woman living in Astoria, Queens, in the state of New York. She resides alone in an apartment on the upper floor of a brownstone home, which she used to share with her late husband who passed away from lung cancer a few years ago. Mary spent her career as a teller at a nearby bank. Since retiring in her late 60s, she has found her days mostly filled with sedentary activities. She has developed a routine that revolves around watching her favorite television shows and catching up on the news. Mary is aware of the importance of maintaining her health and struggles with managing hypertension and diabetes. Her doctor has advised her that increasing her physical activity could provide numerous benefits. Despite this knowledge, she feels overwhelmed and uncertain about how to incorporate exercise into her daily routine.

Mary has considered taking up walking as a low-impact form of exercise, but she is hesitant due to the unevenness of the sidewalks in her neighborhood. The thought of tripping and falling causes anxiety. She also recognizes the heightened risk of slipping on ice during the cold winter months, which are often harsh in New York. Additionally, the sweltering heat of the summer poses another challenge, making it difficult for her to venture outside for extended periods of time. As a resident of an upper-floor apartment, she must navigate a steep and narrow staircase to exit her home, and this has become increasingly challenging for her over the years.

Although Mary wants to adopt a healthier lifestyle and become more active, she faces considerable barriers in her pursuit of increased physical activity. The combination of her living situation, neighborhood conditions, and weather challenges has created a complex web of obstacles that make it difficult for her to find safe and accessible ways to exercise. Nevertheless, Mary remains determined to find a solution that allows her to enjoy a more active lifestyle while accounting for her unique circumstances.

**Scenario**: Mary is eager to incorporate more physical activity into her life, but she faces a vicious cycle. Her sedentary lifestyle has led to mobility challenges and increased frailty, which have further limited her opportunities to engage in physical activity. This downward spiral has left Mary feeling trapped and frustrated as she struggles to find suitable ways to stay active and healthy. To break free from this cycle, Mary tried a video-based exercise routine at home. However, she quickly discovered that these routines did not provide the type of engagement she needed to stay motivated. Mary found the exercises monotonous, and the lack of feedback made it difficult for her to maintain interest and gauge her progress. After only a few weeks, Mary quit.

**Potential Solutions**: VR- and AR-based fitness programs hold great potential for supporting Mary's goal of becoming more physically active as these programs can effectively address many of the barriers she has faced. Like video-based solutions, by offering the convenience of exercising within her own home, these programs eliminate concerns about weather and poorly maintained sidewalks, helping to promote a safe and comfortable environment for Mary to work out in. In such a program, for example, Mary might put on a VR headset and be presented with a suite of exergames to choose from and a virtual coach to instruct her on how to engage in these activities. An exergame is an interactive videogame that combines physical exercise and motivating gameplay elements. Scores, game achievements, and various data visualizations have the potential to allow Mary to easily track and monitor her fitness progress. These metrics might also be used to dynamically adjust the difficulty of the exergames based on her level of performance, creating a tailored and personalized fitness experience. Unlike the exercise videos she tried before, VR and AR programs can facilitate long-term adherence through dynamic, immersive, and engaging environments and gameplay. These programs also have the potential to include motivating social components, either by allowing Mary to exercise in the same virtual space as others who might be distally located, or through leaderboards allowing Mary and her friends and family to compare achievements and encourage one another.

Overall, this type of fitness programs could offer a comprehensive solution to help Mary overcome challenges and achieve her goal of leading a more active and healthier lifestyle. Mary could also benefit from activities that focus on her mental health and wellbeing, especially since the loss of her husband and due to her anxiety about her health. XR meditation applications provide opportunities to assist with this goal as well, offering real-time support for engagement in a variety of relaxation and meditative exercises. A growing body of evidence supports the feasibility, acceptability, and initial efficacy of healthy lifestyle and wellness XR solutions.

## 4.4 XR Solutions for Supporting Healthy Lifestyles and Wellness

There is a growing literature examining the potential of exergames to improve the fitness and function of older adults (Ismail et al., 2022). Several studies at this point have begun to examine the potential of XR solutions to help achieve these aims. Recently, a randomized controlled trial involving 60 participants from an active aging center was conducted to study the impact of VR exercise sessions on their functional fitness and quality of life (Barsasella et al., 2021). Participants were divided into two groups: the intervention group received VR sessions twice a week for 6 weeks, while the control group received no treatment. VR sessions included experience with a variety of VR applications with different movement intensities, from sports experiences that involved high-intensity movements to exploration-based experiences that involved very low-intensity movements. Functional fitness was evaluated through tests administered at the beginning and end of the study. Quality of life and happiness were assessed through questionnaires.

Performance on the back scratch test, a standard assessment of physical functioning that measures how closely the hands can be brought together behind one's back, significantly improved in the intervention group. Additionally, the results indicated a significant increase in happiness in the intervention group compared to the control group. The researchers concluded that VR sessions have the potential to enhance wellbeing and functional fitness in older adults, and the findings are supportive of future research efforts to explore the physical and psychological benefits of XR content. A more recent review specifically explored the effects of VR interventions on physical function, balance, and falls in older adults with balance impairment (Ren et al., 2023). This review of 23 studies found that VR interventions had benefits for this population, and especially benefited older adults in hospital and nursing home settings.

A critical question is whether XR exergames can deliver a workout intense enough to drive other positive health outcomes, such as improved cardiovascular function in older adults. Initial data are positive. In a small study of 22 participants over the age of 65 years, Vorwerg-Gall et al. (2023) had participants engage in both VR strength training and VR endurance training (25 min each), directed by a virtual coach. All participants had hypertension. VR exercises included a basketball-like ball game and dancing, as well as more traditional exercises such as squats and overhead presses. For both strength training and endurance training sessions, changes in blood pressure and heart rate, as well as perceived exertion, were consistent with benchmarks for blood pressure reduction. Thus, it is possible for XR solutions to provide activities that are intense enough to impact important health outcomes.

Turning to the topic of meditation, at present there is not a large body of work examining meditation and relaxation XR solutions to promote health and wellbeing among older adults. However, several pieces of evidence point to the promise of this approach. A recent narrative systematic review collected and evaluated studies that examined the impact of immersive VR-based mindfulness training on adults' (not older adults' specifically) mental health (Ma et al., 2022). Seven studies published from 2019 to 2021 met the inclusion criteria, the majority of which used 3D head-mounted VR headsets. In addition to improving mindfulness, there was evidence that these interventions could help alleviate anxiety and depression, enhance sleep quality, improve mood, and promote emotion regulation.

With respect to understanding XR implementation in this domain, Dilanchian et al. (2021) gave younger and older adults experience with a variety of immersive VR applications through a headset, including experience with a commercial meditation app. This meditation app placed users into a realistic tropical waterfall environment that featured soothing nature sounds and music. Acceptability and feasibility were established across all apps studied. Older adults rated the mental and physical demands of the meditation app as low, and younger adults and older adults did not differ in their ratings of the usability of the app. This was despite older adults having less overall technology proficiency. Surprisingly, in this study, older adults reported better VR experiences compared to younger adults. Older adults felt more present in the virtual environments and reported fewer symptoms of cybersickness.

VR relaxation and meditation interventions offer the opportunity to provide direct guidance, for example through visual and auditory feedback related to breathing exercises, while also placing users into soothing environments. Influential theories have posited that exposure to natural or green spaces might improve mood and reduce stress. In our own research, we explored VR as a tool to improve the mood states of older people with and without cognitive impairment by providing interactive nature-based content (Kalantari et al., 2022). If VR interventions can boost mood, reduce stress, and potentially improve cognition, individuals in the early stages of cognitive impairment might especially benefit from them. We developed a VR environment that combined 360-degree videos of natural areas with interactive features, including a virtual flower garden, allowing users to engage with the environment (Box 4.1). Fifty older people participated, and changes in mood states and attitudes toward VR were measured before and after the VR sessions. Results showed significant improvements in positive mood and increased calm after VR exposure. There were no significant differences between participants with and without cognitive impairments. Although attitudinal barriers are often discussed as to why older adults may not adopt technology as readily as younger adults, older adults' attitudes toward VR were high initially, and improved significantly after VR exposure. These findings highlight the potential of VR as a method to enhance the wellbeing of older adults, especially in situations where access to nature is limited, as in the case of our persona Mary.

## 4.5 Key Initiative

Like many of the XR applications discussed in this book, these healthy lifestyle and wellness interventions are promising with respect to their potential to support the needs of older adults with and without cognitive impairments, but additional studies are required that are larger and more rigorous (featuring strong control groups, and random assignment) to be able to make definitive recommendations. As with many emerging interventions involving technology, these larger studies with carefully chosen measures and control groups are important to better understand not just whether these interventions work, but for whom, and why.

Much more development effort is necessary, taking advantage of user-centered iterative design approaches, to ensure that these solutions are useful and usable by older adults, as well as safe. Due to the risk of injury, safety (discussed in more detail in Chap. 9) is an especially salient issue for interventions that involve headsets that might observe vision while the user is engaged in high-intensity physical activity. Input from healthcare providers, exercise experts, mental health professionals, and older adults themselves is especially crucial in this domain. Guidance by medical professionals and exercise physiologists with expertise in aging can help ensure that, for example, VR and AR exergames provide the right level of challenge to ensure exercise goals are being met while not exceeding safe levels of physical activity. Mental health professionals can also guide the development of XR experiences that are most likely to improve mental wellbeing. Social, health, and sports psychologists can help develop guidance for interactions with virtual wellness coaches (Box 4.2). As always, older adults need to be involved in the design process at all stages of intervention development.

**Box 4.1 Testing a VR Application to Boost Mood**

Kalantari and colleagues developed the VR Garden program, implemented using the Meta Quest 2 headset, to create a virtual nature experience that encourages active participation (Fig. 4.2). Based on therapeutic garden guidelines, older adults could walk through the garden, observing trees, plants, flowers, ponds, and fountains. Interactions included touching flowers to release butterflies, feeding ducks, and throwing rocks into the pond. Handheld controllers provided vibration feedback, while natural sound recordings enhanced realism. The gardening game within the VR Garden allowed older adults with and without cognitive impairments to design garden layouts and plant and water flowers with the aim of promoting engagement. After exposure to the VR Garden, participants demonstrated significant improvements in mood and increased calm. Cybersickness was minimal. Older adults who participated provided a wealth of information related to how to improve and adapt the program to better meet their needs and preferences, contributing to a user-centered, iterative design approach.

**Fig. 4.2** Screenshots from the VR Garden experience, including images of **a** VR training; participants were trained on how to navigate the virtual garden—left, and how to plant virtual flowers—right; **b** 360-degree nature videos; the image here is distorted because it presents the 3D view in a 2D picture; **c** garden interactions in which players interacted with plants, animals, and objects in the virtual environment; and **d** the gardening game in which players designed and planted their own garden

**Box 4.2 Virtual Coaches to Motivate Engagement**

XR platforms to promote engagement in activities that benefit physical and mental health have the potential to do so through virtual coaches. Virtual coaches can provide feedback, help older adults set and maintain realistic goals, and instruct them on how to engage in different activities. What should these avatars look like? And how should they behave? Additional research, with a participatory design focus, is needed to help answer these questions. This research should include interviews and focus group studies, as well as directly obtain feedback from older adults after interactions with prototype coaches. Allowing customization and individualization of virtual coaches can help account for individual differences and diversity among the older adult population. Recent leaps in generative artificial intelligence might allow virtual coaches to interact more naturally with older adults through new developments with respect to the understanding of and production of natural speech.

## References

Barsasella, D., Liu, M. F., Malwade, S., Galvin, C. J., Dhar, E., Chang, C. C., & Syed-Abdul, S. (2021). Effects of virtual reality sessions on the quality of life, happiness, and functional fitness among the older people: A randomized controlled trial from Taiwan. *Computer Methods and Programs in Biomedicine, 200*, 105892.

Centers for Disease Control and Prevention (n.d.). Retrieved from 01 May 2023: https://www.cdc.gov/physicalactivity/basics/older_adults/index.htm

Dilanchian, A. T., Andringa, R., & Boot, W. R. (2021). A pilot study exploring age differences in presence, workload, and cybersickness in the experience of immersive virtual reality environments. *Frontiers in Virtual Reality, 2*, 736793.

Ismail, N. A., Hashim, H. A., & Ahmad Yusof, H. (2022). Physical activity and exergames among older adults: A scoping review. *Games for Health Journal, 11*(1), 1–17.

Kalantari, S., Bill Xu, T., Mostafavi, A., Lee, A., Barankevich, R., Boot, W. R., & Czaja, S. J. (2022). Using a nature-based virtual reality environment for improving mood states and cognitive engagement in older adults: A mixed-method feasibility study. *Innovation in Aging*, 6(3), igac015.

Ma, J., Zhao, D., Xu, N., & Yang, J. (2022). The effectiveness of immersive virtual reality (VR) based mindfulness training on improvement mental-health in adults: A narrative systematic review. *Explore.*

National Institute on Aging (n.d.). Retrieved on 05 May 2023 from: https://www.nia.nih.gov/health/cognitive-health-and-older-adults#stress

Prenderville, J. A., Kennedy, P. J., Dinan, T. G., & Cryan, J. F. (2015). Adding fuel to the fire: The impact of stress on the ageing brain. *Trends in Neurosciences, 38*(1), 13–25.

Ren, Y., Lin, C., Zhou, Q., Yingyuan, Z., Wang, G., & Lu, A. (2023). Effectiveness of virtual reality games in improving physical function, balance and reducing falls in balance-impaired older adults: A Systematic review and meta-analysis. *Archives of Gerontology and Geriatrics*, 104924.

Vorwerg-Gall, S., Stamm, O., & Haink, M. (2023). Virtual reality exergame in older patients with hypertension: A preliminary study to determine load intensity and blood pressure. *BMC Geriatrics, 23*(1), 527.

# 5 Cognitive Health and Dementia

## 5.1 The Challenge

Cognitive changes are a natural part of the aging process (Institute of Medicine, 2015). As we age, we can all expect to experience some degree of decline in abilities such as memory, processing speed, attention, and decision-making. These changes can interfere with our ability to perform everyday tasks, making them more challenging and error-prone. For example, due to declines in memory, an older adult might forget to purchase an ingredient when shopping for dinner. While some cognitive errors result in relatively minor inconvenience (e.g., a return trip to the store), in domains such as medication and financial management, as well as driving, these errors can have serious consequences for an individual's health, safety, and wellbeing. Cognitive aging can also impair an individual's ability to engage in activities that bring joy and a sense of purpose, such as hobbies and volunteer work. Fortunately, older adults acquire vast amounts of knowledge and learned strategies over their lives that can compensate for many cognitive ability changes. Importantly, there is substantial variability in cognitive decline among older adults as a natural part of the aging process, as well as variability in their ability to manage the impacts of these declines. Novel solutions are needed to assist older adults in navigating cognitive changes inherent to the aging process and the consequences of these changes on everyday activities.

Apart from normative age-related changes in cognition, many older people will experience cognitive decline due to injury or disease. In the United States alone, it is estimated that the number of people living with Alzheimer's disease (AD) will increase from around 6 million now to almost 15 million by 2060 (Rajan et al., 2021). Dementia, often thought to result from the accumulation of abnormal proteins in the brain or vascular disturbances, causes much more severe cognitive impairments as compared to typical cognitive aging. These impairments pose a serious threat to an individual's ability to live independently



W. R. Boot et al., *Extended Reality Solutions to Support Older Adults*, Synthesis Lectures on Technology and Health, https://doi.org/10.1007/978-3-031-69220-8_5

and can result in the need for caregiving support and require the transition from an independent living setting to an assisted living setting. Within the home, support is often provided by an informal caregiver (e.g., spouse, adult child), but the demands of caregiving can be substantial and challenging for informal care providers to manage, resulting in stress and burden.

Unless a cure or preventative intervention is found as the population ages, the number of individuals experiencing dementia will increase substantially over the next several decades, placing enormous strain on healthcare and economic systems. Developing successful interventions to address the large and growing need for cognitive support for older adults with and without cognitive impairments is a critical goal, and, as this chapter outlines, XR offers several promising solutions.

## 5.2 What's in this Chapter?

This chapter reviews some of the challenges related to maintaining cognitive health and counteracting the impact of age-related cognitive changes and dementia. The potential of augmented reality (AR) and virtual reality (VR) to address these challenges is discussed, as well as a brief review of the literature discussing the promise and pitfalls of these approaches. These issues are discussed in the context of the persona Billy, an older man concerned about his recent diagnosis of mild cognitive impairment (MCI) who wants to explore ways to maintain cognitive health (Fig. 5.1).

## 5.3 Persona and Scenario

**Persona**: Billy is an 80-year-old man residing in the suburbs of Toronto, Canada. A widower for the past seven years, he lives alone but is generally happy with his daily routine. He passes the time reading, emailing friends and family members, and when the weather is nice, he enjoys golfing. He attends classes at the local senior center and is generally content with his active leisure and social life. However, in recent months, Billy has found himself grappling with increasing difficulties related to his memory and planning abilities. A retired mathematics professor, Billy is known in his social circles for his quick wit and engaging conversation. However, his growing forgetfulness has raised concerns among his friends. Billy is also experiencing difficulties coming up with the right word. More and more often, he misplaces items within his home, like his glasses. After missing a social event he was looking forward to after misremembering the date, Billy decided to seek the advice of a medical professional. Billy's doctor referred him to a neurologist who, after a series of cognitive tests and evaluations, diagnosed him with mild cognitive impairment (MCI). He was told that this condition often acts as an intermediate stage between the normal cognitive decline associated with aging and more

**Fig. 5.1** Billy was recently diagnosed with mild cognitive impairment (MCI) and is looking for novel, mentally stimulating activities to help support his cognition (Shutterstock).

serious neurodegenerative diseases like Alzheimer's disease. Although his challenges are relatively minor now, he worries about the future and is interested in finding ways to help him cope with his cognitive challenges and prevent further decline.

**Scenario**: Like many older people facing cognitive decline, Billy is worried and is motivated to do what he can to stay mentally sharp. He has always valued his independence and is concerned he may have to rely on his daughter, his only child, for assistance or move to an assisted living facility if his cognitive abilities decline further. After consulting with his physician and neurologist, and researching different options himself, Billy learned that there are few successful pharmacological treatments available for cognitive decline associated with MCI or dementia. In his research, however, he did find some scientific studies highlighting behavioral and lifestyle interventions as approaches that have demonstrated some promise, including interventions involving computerized cognitive training. Billy is looking for fun, engaging, and novel cognitive training activities that he can incorporate into his life to help keep his mind sharp. He hopes that this additional cognitive stimulation might help stave off further decline, allowing him to maintain his independence and continue to engage in the activities he enjoys.

**Potential Solutions**: VR- and AR-based applications are currently being investigated with respect to their ability to improve cognition. Billy learned of such a study being conducted at the university where he used to teach. After reading more about the study's requirements, including that it was recruiting individuals with MCI, he agreed to volunteer as a participant. As part of the study, he was asked to perform a series of cognitive assessments, and then was given a VR headset to take home with him. The headset allowed him to engage in several cognitive training activities within exciting and immersive virtual worlds. To Billy, the VR activities felt more like games than training. Over the course of the study, he found that the training activities adapted to his skill level. This kept the training from becoming too easy and thus boring, or too challenging and thus stressful. After 6 months, the researchers tested his cognitive abilities again. When the study is complete, the scientists will determine whether individuals receiving the VR treatment improved more on the cognitive assessments than those in the control condition. As compensation for his participation in the study, Billy was allowed to keep the headset and training software. Even though his participation is over, he continues to engage in the VR training. Billy feels less anxious and more in control knowing that he is engaging in activities that challenge his mind, and he finds the training activities engaging and enjoyable.

## 5.4 XR Solutions for Supporting Cognition and Cognitive Engagement

So far, there is mixed evidence regarding the efficacy of traditional cognitive training programs to improve the performance of important everyday tasks (Nguyen et al., 2022; Simons et al., 2016). The central question in this literature is whether performance gains within the context of cognitive training activities translate to performance improvements on important everyday tasks, such as driving and medication management. Despite this uncertainty, XR is likely to play important roles in helping older adults maintain their cognition and independence. First, there are several reasons to believe that the efficacy of cognitive training programs could benefit from the more immersive experience that XR can provide. In many XR application domains, an individual's sense of presence (the sense of "being there" in the virtual world) appears to be critical for intervention success and promoting better learning outcomes. Immersive VR, by facilitating presence, has the potential to boost the effectiveness of cognitive training activities compared to typical, less immersive computerized cognitive training. Second, transfer of training gains is thought to depend on the similarity between the training task and environment and the target task and environment. Ultimately, it is of little practical significance that older adults improve in their performance of cognitive training activities. The outcome of interest is improved performance on everyday tasks that matter to them and their independence. Immersive VR has the potential to create training tasks and environments much more like the "real world," maximizing the likelihood of training gains transferring to everyday tasks. Finally, the engaging and compelling nature of XR has the potential to boost training adherence, which can be low in cognitive training programs. Effective cognitive training programs that don't inspire frequent engagement will not meet their potential.

Initial studies have indicated some promise with respect to the potential of cognitive training with VR to boost cognitive abilities. Wais et al. (2021), for example, developed a VR training program for older adults that capitalized on the overlapping neural circuitry responsible for spatial navigation and memory within a region of the brain called the hippocampus. The training involved participants engaging in a variety of complex wayfinding tasks in immersive virtual environments. Participants moved through these environments by walking in places; motion tracking devices detected ankle movements, and this was translated into forward motion in the virtual environment. The study, similar to the study Billy participated in, assessed memory before and after participants received 12 hours of this virtual wayfinding game training or training on a control game presented on a tablet computer. Relative to the control group, the VR group demonstrated significant gains in memory performance, and memory gains were correlated with improvements in the VR game. The researchers attributed the success of their approach in part to the way VR can engender a high degree of engagement, and the ability of immersive VR to activate navigation circuits in the brain like the way in which these circuits are engaged during real-world navigation and wayfinding tasks.

With respect to older adults with cognitive impairment specifically, a review of seventeen randomized controlled trials (RCTs), involving 744 participants, found evidence that VR cognitive training had significant benefits on measures of global cognitive and executive function in patients with MCI (Zhong et al., 2021). However, the meta-analysis also showed no effects of VR cognitive training on delayed memory, immediate memory, attention, and everyday activities. While results appear promising with respect to the ability of VR to enhance some cognitive functions, like all meta-analyses, confidence in the results depends in part on the quality of the studies included in the analysis. Some studies had small sample sizes, harming the ability of these studies to detect small cognitive training effects. Some studies compared individuals receiving VR cognitive training with control groups that received no training at all. These studies cannot fully rule out placebo effects as being responsible for the observed cognitive training effects. Placebo effects refer to improvements not due to the intervention itself, but to participants' expectations that they should improve because they have undergone a treatment. Many studies as well were of limited duration, raising questions of whether larger doses of VR might have produced stronger results, and of how long-lasting benefits might be.

Beyond mild cognitive impairment, few studies have examined the impact of XR cognitive training on cognitive abilities among people with more serious neurocognitive conditions or disorders. Moreno et al. (2019) conducted a review of VR studies and found only three studies that included participants with Alzheimer's disease, and only six studies that included participants who had experienced a stroke. Furthermore, the vast majority of studies uncovered (82%) did not involve fully immersive VR experiences (i.e., many training programs displayed the virtual environment on a screen or computer monitor).

Thus, the potential of fully immersive VR experiences to address the cognitive challenges of individuals with acquired brain injury or dementia remains largely unknown.

Rather than train cognitive abilities in the hopes that training gains will translate to better performance on everyday tasks, an alternative approach is to train or support cognitively demanding everyday tasks directly through XR. Outside of the context of XR, Czaja et al. (2020) demonstrated that this approach can be quite successful. They trained older adults with and without cognitive impairment using ecologically valid simulations of financial management, medication management, and transportation tasks on a touch-screen computer. Training resulted in substantial performance gains on these tasks, and gains did not differ depending on cognitive status. XR has the potential to expand upon this successful approach. In the context of XR, additional, complex everyday tasks might be trained with higher fidelity, helping to ensure that the training on simulated tasks translates to better performance outside of XR. Perhaps more exciting, AR guidance has the potential to provide individuals, including older adults, with real-time guidance for how to perform complex, everyday tasks even in the face of substantial cognitive decline, boosting their ability to live independently longer (Blattgerste et al., 2019). In the future, for example, visual cues superimposed over the kitchen environment, combined with auditory instruction, might help older adults experiencing cognitive impairment cook nutritious and delicious meals. These solutions have tremendous potential to assist someone like Billy perform tasks required for independent living should his cognition continue to deteriorate over time.

In addition to supporting cognition and cognitively demanding activities, XR also shows promise with respect to assessing cognition and detecting cognitive change (Box 5.1). Once cognitive changes are detected, XR solutions might be recommended to counteract these changes or support the performance of important everyday tasks in spite of these changes.

## 5.5 Key Initiative

Cognitive aging and age-related cognitive decline are urgent challenges that require novel, creative solutions, and XR is likely to play an important role in these efforts. Ongoing and upcoming clinical trials are exploring the potential benefits of XR interventions on cognitive health and the performance of instrumental activities of daily living. A few studies conducted so far have demonstrated acceptability, feasibility, and initial efficacy of this approach. However, large-scale studies with diverse older adult participants, as well as longer follow-up periods, are required to truly understand this approach's potential. An alternative, and a potentially more fruitful, method is to use XR to directly train older adults' performance of important everyday tasks. This approach is promising because it relies less on "far transfer." That is, there is a high correspondence between the training and target task, making training gains likely to transfer outside of XR. Which approach, or whether a combination of both approaches, might be more successful in helping older adults remain independent is an important research initiative.

With respect to AR solutions to provide real-time support for the performance of cognitively demanding tasks, this approach also has tremendous potential. These AR approaches may be most beneficial to individuals already experiencing significant cognitive impairment. However, the development of such systems is complex, specifically with respect to computer vision problems of recognizing relevant objects in the environment for the task, the task steps that have been completed so far, and potential errors with enough time to intervene. Another important question is how best to provide online guidance, through visual and auditory cues to ensure that tasks can be completed quickly and with minimal error while also not interfering with the processing of visual and auditory components of the task.

**Box 5.1 XR to Assess Cognition**

In addition to bolstering the performance of cognitively demanding activities, extended reality (XR) can be used to assess cognition and detect cognitive decline. For instance, Parson and Rizzo (2008) developed the Virtual Reality Cognitive Performance Assessment Test (VRCPAT), which incorporates a memory test within the exploration of a 3D virtual city environment. This test was administered to older adults, in conjunction with traditional measures of cognitive ability. As expected, higher VRCPAT scores were associated with higher scores on established measures of memory. However, there was no association between VRCPAT scores and measures of attention, processing speed, executive function, or verbal fluency, which supports VRCPAT's validity as a specific measure of memory capability. One of the benefits of VR-based assessments is their enhanced ecological validity. Since participants are evaluated using tasks and stimuli that closely mimic real-world conditions, as opposed to traditional (frequently paper-and-pencil) methods, it is more likely they tap into cognitive mechanisms and strategies pertinent to daily performance. Furthermore, VR yields a wealth of detailed data to characterize behavior, including motion tracking data as participants interact with the virtual environment. These data hold the potential to offer nuanced, indirect measures of cognition and cognitive change. The engaging nature of VR could also foster frequent testing, including self-assessment, among older adults. Tests of this nature could ultimately prove to be more sensitive to early cognitive impairment compared to traditional measures, as posited by Jang et al. (2023).

## References

Blattgerste, J., Renner, P., & Pfeiffer, T. (2019). Augmented reality action assistance and learning for cognitively impaired people: a systematic literature review. In *Proceedings of the 12th ACM international conference on pervasive technologies related to assistive environments* (pp. 270–279).

Czaja, S. J., Kallestrup, P., & Harvey, P. D. (2020). Evaluation of a novel technology-based program designed to assess and train everyday skills in older adults. *Innovation in Aging, 4*(6), 1–10.

Institute of Medicine. (2015). *Cognitive aging: Progress in understanding and opportunities for action.* The National Academies Press.

Jang, S., Choi, S. W., Son, S. J., Oh, J., Ha, J., Kim, W. J., & Seok, J. H. (2023). Virtual reality-based monitoring test for MCI: A multicenter feasibility study. *Frontiers in Psychiatry, 13*, 1057513.

Moreno, A., Wall, K. J., Thangavelu, K., Craven, L., Ward, E., & Dissanayaka, N. N. (2019). A systematic review of the use of virtual reality and its effects on cognition in individuals with neurocognitive disorders. *Alzheimer's & Dementia: Translational Research & Clinical Interventions, 5*, 834–850.

Nguyen, L., Murphy, K., & Andrews, G. (2022). A game a day keeps cognitive decline away? A systematic review and meta-analysis of commercially-available brain training programs in healthy and cognitively impaired older adults. *Neuropsychology Review, 32*(4), 601–630. https://doi.org/10.1007/s11065-021-09515-2

Parsons, T. D., & Rizzo, A. A. (2008). Initial validation of a virtual environment for assessment of memory functioning: Virtual reality cognitive performance assessment test. *Cyber Psychology & Behavior, 11*(1), 17–25.

Rajan, K. B., Weuve, J., Barnes, L. L., McAninch, E. A., Wilson, R. S., & Evans, D. A. (2021). Population estimate of people with clinical Alzheimer's disease and mild cognitive impairment in the United States (2020–2060). *Alzheimer's & Dementia.* https://doi.org/10.1002/alz.12362

Simons, D. J., Boot, W. R., Charness, N., Gathercole, S. E., Chabris, C. F., Hambrick, D. Z., & Stine-Morrow, E. A. (2016). Do "brain-training" programs work? *Psychological Science in the Public Interest, 17*(3), 103–186.

Wais, P. E., Arioli, M., Anguera-Singla, R., & Gazzaley, A. (2021). Virtual reality video game improves high-fidelity memory in older adults. *Scientific Reports, 11*(1), 2552.

Zhong, D., Chen, L., Feng, Y., Song, R., Huang, L., Liu, J., & Zhang, L. (2021). Effects of virtual reality cognitive training in individuals with mild cognitive impairment: a systematic review and meta-analysis. *International Journal of Geriatric Psychiatry, 36*(12), 1829–1847.

# 6 Getting Around

## 6.1 The Challenge

This chapter focuses on the act of getting from one place to another, also known as transportation. Tasks that are included under the transportation category can vary drastically in their complexity and scale, from moving to a different room within a familiar residential environment to walking to a local park, to locating a specific clothing store in the labyrinth of a crowded shopping mall, to navigating through an airport or train station, to embarking on a cross-country road trip to a new city. Transportation activities can be affected by a variety of age-related changes, both physical and cognitive, which may reduce older adults' ability to move freely and easily from one place to another.

Transportation is considered an instrumental activity of daily living (IADL). In other words, it is a crucial aspect of being able to live independently, with a similar importance as activities such as cooking, shopping, managing one's medications, and performing housekeeping tasks. Transportation is regarded as a "keystone" activity for successful aging because it is required for the completion of many other important everyday tasks. It can be much more challenging for an older adult to manage health needs if they cannot easily get to a doctor's office or pharmacy, and it can be difficult to cook meals and engage in housekeeping if one cannot easily get to shops to buy food and cleaning supplies. Transportation also allows older people to enjoy hobbies and leisure activities, such as going to parks or events and visiting with friends and family. For many older adults, transportation is integral to their independence, wellbeing, and quality of life.

Wayfinding, the cognitive process through which we navigate and understand our surroundings, is central to numerous transportation activities. Unfortunately, some older adults face challenges when it comes to wayfinding. Age-related changes in vision can make reading navigational signs more difficult, whether in a mall or on a highway. In



W. R. Boot et al., *Extended Reality Solutions to Support Older Adults*, Synthesis Lectures on Technology and Health, https://doi.org/10.1007/978-3-031-69220-8_6

addition, spatial cognition—the ability to understand, mentally manipulate, and remember spatial relationships between objects and locations—often deteriorates as we age (Klencklen et al., 2012). Declines in spatial abilities can be especially pronounced for older adults who have a cognitive impairment, which can range from mild cognitive impairment (MCI) to the effects of dementia, Alzheimer's disease, traumatic brain injury, or stroke (DeIpolyi et al., 2007; Plácido et al., 2022). Depending on the type and level of impairment, disorientation can occur even in previously familiar surroundings.

The repercussions of wayfinding difficulties can span from minor inconveniences to severe safety concerns. When an individual loses their way and arrives late for an appointment, they may experience embarrassment and frustration. This may lead some older adults to withdraw from activities they enjoy outside their homes if they lose confidence in their wayfinding abilities, contributing to social isolation and loneliness (see Chap. 3). In more severe cases, a person might find themselves dangerously lost, risking exposure to harsh weather conditions, traffic accidents, or other physical dangers. Supporting wayfinding and other transportation needs is therefore crucial for promoting the safety, wellbeing, confidence, and independence of older adults. Given the importance of wayfinding, it is exciting that new technologies such as extended reality (XR) can provide innovative solutions to these challenges.

## 6.2 What's in This Chapter?

This chapter reviews some of the transportation challenges that are caused by age-related changes in wayfinding abilities. The discussion here does not extend to physical impairments that reduce mobility, but instead considers technologies that can assist in wayfinding's cognitive aspects. The potential of augmented reality (AR) and virtual reality (VR) to address these challenges is discussed, along with the potential pitfalls of these approaches. The relevant issues are framed in the context of the persona Annette, an older woman living in a large city who used to walk everywhere but is now no longer confident in her wayfinding abilities due to the cognitive challenges she experienced after a stroke (Fig. 6.1).

**Fig. 6.1** Annette has withdrawn from activities outside of her home because of wayfinding challenges related to a recent stroke (Shutterstock)

## 6.3 Persona and Scenario

**Persona**: Annette, an 81-year-old woman, resides in London with her 79-year-old husband, Raymond. They have enjoyed many years in their neighborhood, where almost everything she needs is within walking distance. Her daily routine includes strolls through a nearby park for exercise, and a few times a week she walks to the grocery store to pick up fresh ingredients for dinner. This routine has kept her both active and in touch with her community. However, one day, Annette awoke in a hospital, confused, and faced with the news that she had experienced a stroke. As a result of the injury to her brain, she encountered moderate cognitive challenges. While these challenges lessened over time, some cognitive problems persisted even months later, leading to significant changes in her lifestyle. Despite being physically able to maintain her routine, Annette faces challenges when going places due to her cognitive wayfinding difficulties. She fears getting lost in her own neighborhood and being unable to find her way back home. Navigating unfamiliar places poses even more substantial challenges. The fear of becoming lost has introduced stress into her everyday activities, robbing her of the pleasure she once found in them. She now relies on Raymond to do many of the shopping tasks that she used to perform. The loss of autonomy and independence she has experienced post stroke has stirred anxiety and given rise to feelings of depression.

**Scenario**: Annette yearns to return to her old routine without worrying about getting lost, but her confidence in her wayfinding ability has been shaken. She has attempted to use various GPS apps as a solution; however, deciphering the maps and directions on the small screen of her smartphone, including the unfamiliar symbols that must be mentally transposed onto landmarks in the real environment, has been challenging. As she puts it, she has a hard time "getting the map into my head." She doesn't trust herself not to get lost, even with the aid of GPS apps.

**Potential Solutions**: VR- and AR-based wayfinding programs have significant potential for supporting individuals' navigation abilities. Fortunately, Annette has recently discovered AR-enabled smartphone navigation apps through her tech-savvy daughter. Her phone already had a navigation app on it with an AR option for her city. This app uses the smartphone's camera to superimpose virtual elements onto real-time images of the surrounding area, making it much easier for Annette to connect the navigational guidance to her real environment. After completing a tutorial and becoming familiar with the app, Annette decided to meet a friend at a new local restaurant for lunch. She entered the address of the restaurant into the app, and then once she was on the street she scanned her surroundings with her phone's camera. She saw that the app allowed her phone to act like a window she could look through. The app superimposed a large blue arrow pointing down the street, indicating the direction of travel to reach the address. Above this arrow was the instruction to walk 150 feet. She lowered her phone and walked in that direction. Upon reaching the next intersection, her phone vibrated, and she raised it again and saw an arrow pointing to the left. After quickly reaching the street where the restaurant was located, Annette lifted her phone once more, revealing a red marker at the entrance, with the restaurant's name superimposed on the screen. With a clear understanding of her destination, Annette put her phone away. After lunch, she was easily able to find her way home using the same approach. After several positive experiences using the app, Annette regained her confidence in navigating both familiar and unfamiliar places and being able to find her way back home.

## 6.4 XR Solutions for Supporting Wayfinding and Transportation Safety

The app that helped Annette to resume her old routine with renewed confidence is only one of the many commercially available XR technologies that provides wayfinding support with a potential to benefit older people. Google Maps offers a "Live View" feature that operates much like the app described in Annette's story. By recognizing landmarks in the environment (based on analyzing images from Google's Street View) and overlaying guidance directly onto the real world, this technology minimizes the human spatial processing effort needed to translate directions into action.

However, this type of app is generally only helpful in reaching a building's exterior. After arriving at a particular address, an older adult may then face the daunting task of navigating through a complex interior environment, for example by finding a specific doctor's office in a large and sprawling medical complex or finding a specific airport terminal. The ability to provide indoor AR wayfinding support is currently limited by the sparse available data on such environments. Several companies are now starting to fill this gap by offering tools to help building managers map their spaces and collect the information needed to provide AR guidance to visitors (for example, www.eyedog.us). Large facilities such as shopping malls, medical centers, and transportation hubs are at the forefront of aiding their older users by offering localized AR navigation support. One day, AR guidance might even help an individual with cognitive decline navigate to the grocery store, and then provide specific guidance within the store from aisle to aisle, helping them to find the items on their grocery list. Younger adults, and individuals without cognitive impairment, may also benefit from such technologies when confronted with confusing navigational tasks; but the advantages of AR-assisted navigation are likely to be most pronounced for those who struggle with everyday wayfinding.

While AR smartphone apps are a promising solution to improve wayfinding independence for older adults, they do give rise to some concerns. One objection that frequently arises is that, by relying on such tools, users may fail to exercise their spatial abilities, leading to technological dependency and greater stress and failure rates when the app is not available or effective for a particular environment. This concern is belied by research showing that well-designed AR-assisted navigation can promote active learning and actually improve spatial navigation skills for some users. The interplay between virtual and physical environments can help to strengthen memory retention and cognitive processing, while also providing positive reinforcement to maintain motivation and self-confidence in learning new routes or places (Huston & Hamburger, 2023).

A more concerning issue with AR apps is the physical safety issues that may result from an excessive focus on the smartphone's screen instead of the surrounding area. When using these apps, it is important to only view them intermittently, preferably from a stationary position, so that they do not distract attention from environmental dangers such as traffic and potential tripping hazards. Even fully visible hazards can sometimes go unnoticed when attention is focused on the screen's AR elements, a phenomenon known as inattentional blindness (Mack & Rock, 1998). Having to frequently raise one's phone to scan the environment for AR instructions can also be awkward and fatiguing for older adults.

Smartphone apps can mitigate this concern by incorporating safety warnings and seeking to identify problematic usage (e.g., a significant period of continued use while moving). Future AR support could also enhance user safety by incorporating links to surrounding smart traffic systems: for example by notifying users when it is unsafe to cross the street (Malik et al., 2023). Advances in AR head-mounted displays and glasses have the potential to further improve safety and convenience, allowing users to focus more on their surroundings and making the process less physically demanding (Zhang et al., 2021; Zhao et al., 2020). Such AR apps, if thoughtfully designed, can potentially make a broad range of information available to users on demand, including points of interest or historical notes as well as navigational guidance, while still leaving the field-of-view clear of distractions throughout most of the transportation journey.

In addition to AR, VR presents an alternative solution for improving the wayfinding abilities of older adults, both with and without cognitive impairments. It is possible to develop a VR support system designed to train individuals to navigate routes within specific indoor spaces, such as hospitals, bus stations, and shopping complexes. This approach has the potential to enhance wayfinding abilities, boost confidence, and reduce anxiety when navigating unfamiliar places. VR has the capability to recreate real locations in virtual form, allowing individuals who face wayfinding challenges or anxiety to practice navigating to a location with guidance and real-time feedback multiple times before they need to navigate to that location in the real world. This method is promising for several reasons. The strong sense of presence or "being there" that VR provides is associated with enhanced learning as compared to traditional screen-based activities (Makransky & Petersen, 2021). Another notable advantage of VR is that it can implement errorless learning methods, in which the technology intervenes to prevent users from completing an incorrect action. This approach reduces the likelihood of errors being consolidated during the learning process and is particularly beneficial for individuals with cognitive impairment.

An initial proof-of-concept study for this method was conducted by Lokka et al. (2018). In this case, researchers used non-immersive video-based virtual cityscapes to provide training for learning routes within a city, and they also investigated the best methods to present virtual environments to maximize learning. They found that both younger and older adults could use this method to learn routes effectively, and that the most effective learning occurred when the VR environment highlighted important landmarks, rendering them with high-definition realism, while using less detailed grayscale textures for other aspects of the environment. Although their studies involved younger adults, Hejtmanek et al. (2020) demonstrated that immersive virtual reality can serve as an effective means to train real-world wayfinding. All participants were tasked with learning the locations of six offices within a building. They learned these locations through one of three methods: physically walking around the building, walking in a place on a treadmill while immersed in a VR headset that displayed a virtual version of the building, or navigating a virtual version on a desktop computer with a mouse and keyboard. Unlike the study of Lokka et al. (2018), this study investigated transfer to the real world. After three practice rounds, all participants attempted to navigate routes within the real building. Although learning that occurred in the real environment was most effective, there was significant transfer of learning demonstrated by participants in the immersive VR and desktop VR conditions, with a slight advantage for the immersive VR condition.

Rather than training specific routes, an alternative might be to exercise cognitive abilities important for navigation and wayfinding (see Chap. 5), or to train wayfinding strategies using VR. Claessen et al. (2016) investigated this approach in a sample of six participants experiencing cognitive impairment post stroke. For example, participants with impaired route knowledge were encouraged during VR training to explore a strategy relying more on survey knowledge. Route knowledge refers to sequential, directional instructions in which landmarks help an individual know what to do next (e.g., turn left past the cafeteria). Survey knowledge, on the other hand, relies more on a map-like representation of the environment, containing information about distances and spatial relationships. A VR training approach was found to be feasible and acceptable, shifting the navigation strategy of most participants. Immersive strategy training programs featuring more adaptive feedback and guidance are likely to be even more effective.

As many of these experiments indicate, wayfinding activities in VR can serve as a valuable research platform, helping us to develop key insights about older adults' navigation abilities and effective navigational aids. Since MCI, dementia, and Alzheimer's disease are strongly associated with spatial deficits (DeIpolyi et al., 2007), navigational testing in VR may someday also assist in diagnosing, or even predicting, these conditions (Box 6.1).

## 6.5 Key Initiative

XR solutions show great promise in their potential to enhance the indoor and outdoor wayfinding abilities of older adults, which could significantly boost their independence and transportation options. Some of these solutions are relatively mature, as is the case for existing AR smartphone apps providing walking directions. Other emerging solutions could soon become commonplace, but are currently faced with technical challenges. For instance, AR headsets and glasses are still often cumbersome and heavy, and are limited by their processing power and battery life. Perhaps most importantly, research development in this area confronts a difficult ethical trade-off when it comes to balancing safety and independence, particularly for older adults with significant cognitive impairments. How can we best promote the autonomy of such individuals, while still helping to ensure that they remain safe? The preferences of older adults for AR versus VR solutions should also be further explored, as well as the effectiveness of both in supporting older adults with diverse cognitive challenges.

There is a clear path forward for research in this area. It involves gaining a better understanding of how the design of XR technologies for navigation can accommodate the diverse needs, preferences, and abilities of older users. The level of support that is required and appropriate for different levels of impairment needs to be better understood, through robust empirical study with diverse older adult populations. In particular, researchers should carefully evaluate how the design of XR navigational tools affects users' ability to attend to environmental hazards, and how to develop systems that can automatically detect and alert users to such hazards. Following this agenda will result in XR-based wayfinding solutions that are not only useful to older adults with diverse abilities, but safe as well.

**Box 6.1 Can Wayfinding in VR Predict Alzheimer's Disease?**
A recent study by Howett et al. (2019) evaluated the VR navigational abilities of patients with mild memory issues, some of whom had brain markers suggesting that they might develop Alzheimer's disease. This task was meant to exercise a specific area of the brain known as the entorhinal cortex, which is key for navigation and is often one of the first brain regions affected by Alzheimer's. The study found that participants who had Alzheimer's disease markers did worse on the navigational task compared to those who did not. The VR navigational task was more successful at differentiating between these two groups than traditional cognitive tests. Given these and other promising results, in the near future XR methods may come to play an important role in the early detection and treatment of Alzheimer's disease and other cognitive impairments.

## References

Claessen, M. H., van der Ham, I. J., Jagersma, E., & Visser-Meily, J. M. (2016). Navigation strategy training using virtual reality in six chronic stroke patients: A novel and explorative approach to the rehabilitation of navigation impairment. *Neuropsychological Rehabilitation, 26*, 822–846.

Delpolyi, A. R., Rankin, K. P., Mucke, L., Miller, B. L., & Gorno-Tempini, M. L. (2007). Spatial cognition and the human navigation network in AD and MCI. *Neurology, 69*(10), 986–997.

Hejtmanek, L., Starrett, M., Ferrer, E., & Ekstrom, A. D. (2020). How much of what we learn in virtual reality transfers to real-world navigation? *Multisensory Research, 33*(4–5), 479–503.

Howett, D., Castegnaro, A., Krzywicka, K., Hagman, J., Marchment, D., Henson, R., & Chan, D. (2019). Differentiation of mild cognitive impairment using an entorhinal cortex-based test of virtual reality navigation. *Brain, 142*(6), 1751–1766.

Huston, V., & Hamburger, K. (2023). Navigation aid use and human wayfinding: How to engage people in active spatial learning. *KI-Künstliche Intelligenz*, 1–8.

Klencklen, G., Després, O., & Dufour, A. (2012). What do we know about aging and spatial cognition? Reviews and perspectives. *Ageing Research Reviews, 11*(1), 123–135.

Lokka, I. E., Çöltekin, A., Wiener, J., Fabrikant, S. I., & Röcke, C. (2018). Virtual environments as memory training devices in navigational tasks for older adults. *Scientific Reports, 8*(1), 10809.

Mack, A., & Rock, I. (1998). *Inattentional blindness*. The MIT Press.

Makransky, G., & Petersen, G. B. (2021). The cognitive affective model of immersive learning (CAMIL): A theoretical research-based model of learning in immersive virtual reality. *Educational Psychology Review*, 1–22.

Malik, J., Kim, N. Y., Parr, M. D., Kearney, J. K., Plumert, J. M., & Rector, K. (2023). Do simulated augmented reality overlays influence street-crossing decisions for non-mobility-impaired older and younger adult pedestrians? *Human Factors, #*, 00187208231151280.

Plácido, J., de Almeida, C. A. B., Ferreira, J. V., de Oliveira Silva, F., Monteiro-Junior, R. S., Tangen, G. G., & Deslandes, A. C. (2022). Spatial navigation in older adults with mild cognitive impairment and dementia: A systematic review and meta-analysis. *Experimental Gerontology, 165*, 111852.

Zhang, J., Xia, X., Liu, R., & Li, N. (2021). Enhancing human indoor cognitive map development and wayfinding performance with immersive augmented reality-based navigation systems. *Advanced Engineering Informatics, 50*, 101432.

Zhao, Y., Kupferstein, E., Rojnirun, H., Findlater, L., & Azenkot, S. (2020). The effectiveness of visual and audio wayfinding guidance on smartglasses for people with low vision. In *Proceedings of the 2020 CHI conference on human factors in computing systems* (pp. 1–14).

# 7 Leisure Activities

## 7.1 The Challenge

In the development and design of technologies to support older adults, there is often a bias toward considering how technology might aid in their performance of instrumental activities of daily living (IADLs). IADLs are defined as activities that one must be able to perform to live independently, such as managing finances, using transportation, shopping for groceries, preparing meals, managing medications, and performing housework. The inability to perform these tasks often results in a reliance on a caregiver or a need to transition from independent to assisted living. While supporting IADLs is an important and urgent goal given the aging of the population, it is also worth recognizing that quality of life and wellbeing are determined by more than one's ability to perform tasks related to independence.

Less attention has traditionally been paid to the design of technologies to support leisure activities and recreational opportunities among older people, which can encompass hobbies, entertainment, recreation, and avenues for new learning. This has been an unfortunate oversight. Evidence suggests that these activities are associated with stress reduction and better physical and emotional health. However, these outcomes are in addition to their primary benefit: providing meaningful, enjoyable, and pleasurable experiences in life. These and other enriching activities have been termed as enhanced activities of daily living (EADLs; Rogers et al., 2020). Given the many advantages EADLs offer to the aging population, designers should be sensitive toward and design technologies to support these needs.

Post retirement, many older adults have more time to devote to leisure pursuits. Younger adults report engaging in four to five hours of leisure activity each day, whereas older adults report about seven hours (U.S. Bureau of Labor Statistics, 2021). Most of this additional leisure time gained later in life is devoted to watching television, by far



W. R. Boot et al., *Extended Reality Solutions to Support Older Adults*, Synthesis Lectures on Technology and Health, https://doi.org/10.1007/978-3-031-69220-8_7

the most common leisure activity of younger and older adults. However, when asked about their favorite leisure activities, older adults tend to report more active and productive pursuits, such as walking, playing sports, arts and crafts, and volunteering (Szanton et al., 2015). Like IADLs, though, age-related changes in health, cognition, and mobility can interfere with the performance of EADLs.

Augmented reality (AR) and virtual reality (VR) solutions hold tremendous potential for promoting participation in meaningful and enjoyable leisure activities later in life. If well-designed, these technology solutions can offer the opportunity to engage in these activities despite substantial age-related changes in health, cognition, and mobility, and in a variety of living contexts (e.g., independent living, assisted living, healthcare settings).

## 7.2 What's in This Chapter?

This chapter reviews some of the challenges related to maintaining activity engagement later in life. The potential of AR and VR to address these challenges is explored, as well as a brief review of the literature examining the promise and pitfalls of these approaches. These issues are discussed in the context of the persona Annie, an older woman who lives in an assisted living facility due to health and cognitive challenges (Fig. 7.1).

**Fig. 7.1** Annie and her family seek a solution that allows her to engage in enjoyable and meaningful leisure activities, despite her significant age-related challenges (Shutterstock)

## 7.3 Persona and Scenario

**Persona**: Annie is an 87-year-old woman who lives in Lawrence, Kansas, in the United States. She recently transitioned from independent living to an assisted living facility due to a constellation of escalating health, mobility, and cognitive challenges that her family found increasingly difficult to manage on their own. Her mobility limitations in particular became severe, and walking even short distances was a challenge for her. Before this transition, when her health and cognition were better, she enjoyed a range of leisure activities at home and in her community. She particularly enjoyed engaging in arts and crafts as a creative outlet, including scrapbooking and making her own greeting cards for friends and family. She also played games with friends, including Mahjong and Bridge. However, since the move, she has faced fewer opportunities to remain engaged and active. While Annie's family visits her as frequently as their schedules allow, she spends many hours of her day in front of the television, passively watching shows without much engagement. Her family wishes for her to participate in more active, intellectually stimulating, and creative activities. However, her health, mobility, and cognitive challenges restrict the range of activities she can safely and enjoyably engage in.

**Scenario**: Annie is inactive and increasingly disengaged. Her family has noticed that, with each visit, she seems more apathetic and depressed. She spends most of her time in her room and is unwilling or unable to participate in many activities organized by the assisted living facility.

**Potential Solutions**: Aware of the challenges Annie faces and her increasing sense of disengagement, the staff at the assisted living facility introduced her family to an innovative VR program they are piloting. This program is designed with residents like Annie in mind, aiming to provide meaningful, engaging, and enjoyable leisure opportunities without the need for physical mobility or even leaving their rooms. With the staff's assistance and using the VR headset, Annie is presented with a host of possibilities. The system offers her fun, simple, engaging games tailored to her cognitive ability. These games do not require her to learn how to use VR controllers; the system tracks her hands, allowing her to interact naturally and intuitively with game elements. Verbal and visual guidance provides instruction and support, helping her learn how to play. Additionally, the program includes virtual arts and crafts sessions. For instance, a virtual painting application allows her to draw using her finger in virtual space. She is grateful for the opportunity to express her creativity again. These drawings can be saved and later shared with family members during their visits. Furthermore, the VR program enables Annie to virtually explore new environments, offering experiences of diverse places, sights, and sounds. She can take a virtual stroll in a serene garden, visit a beach, or even enjoy a virtual symphony concert. These experiences reignite her sense of wonder, curiosity, and connection to the world. These sessions become a topic of discussion during her family's visits as she excitedly shares her virtual adventures and creations with them. The positive impact of the program on Annie is evident to her family. She is more animated and engaged, and her mood is improved.

## 7.4 XR Solutions for Supporting Leisure Activities

By default, many commercial XR systems are leisure-oriented and are meant to provide entertainment. More specifically, most are gaming-oriented, though social, cultural, creative, and new learning opportunities are also popularly downloaded applications. Like many emerging technologies, it is often unclear whether the design of these apps to support leisure activities considers the needs, preferences, and abilities of diverse groups of older adults, including those with physical disabilities or cognitive impairments. However, the potential of XR solutions to support activity engagement across the lifespan, including leisure activities, is beginning to be recognized more broadly. Multiple companies are now marketing specific VR systems aimed at older adult consumers and organizations that serve older people.

Researchers are just beginning to explore the potential of XR solutions for older adults with diminished physical and cognitive abilities, like Annie. For example, Appel et al. (2020) conducted a feasibility study involving sixty-six participants recruited from a long-term care facility, continuing care hospital, community-based memory clinic, or a community-based rehabilitation program. All participants were exposed to a number of nature-based 360-degree videos (e.g., beach and forest scenes) using a head-mounted VR display. Notably, this study examined a relatively low-cost VR option that utilized a smartphone placed within a headset that converted the phone into a head-mounted display. The VR session either took place when the participant was in their bed, their own wheelchair, or in a swivel chair. Encouragingly, headset comfort was rated as high. All participants completed the study, and there were few reported instances of nausea, disorientation, or dizziness (i.e., symptoms of cybersickness). For the most part, positive feelings increased after the VR intervention, though two exceptions were noted. Unexpectedly, participants reported increases in feelings of tiredness and loneliness. Overall, however, participants were able to engage with and enjoy the experience. A major limitation of this study was the short duration of the VR intervention; participants experienced only a single exposure session of 20 min. Brimelow et al. (2020) conducted a highly similar study, exposing thirteen older adults with mild to severe cognitive impairment in a residential care facility to 360-degree videos, using the same VR system. Participants indicated low levels of discomfort, few negative side effects, and that the experience was enjoyable. A significant decrease in apathy was also observed, as measured by facial expression, eye contact, physical engagement, and speech.

Fiocco et al. (2021) conducted a small study to examine potential benefits of VR-supported leisure activities over a longer, multi-week period among older adults within a residential care facility. Specifically, researchers provided residents with the opportunity to engage in "virtual tourism" through a variety of 360-degree videos that lasted between 6 and 10 min, again delivered via a relatively low-cost smartphone-based VR system. Most participants reported traveling frequently in the past, but for a variety of reasons, they no longer engaged in travel (e.g., mobility limitations, lack of financial resources). This intervention allowed older adults to virtually tour a variety of unique international locations. The intervention took place in designated lounge areas of the facilities that participated. Older adults viewed one of these videos three times a week for six weeks. Short-term effects of VR exposure included reductions in anxiety. After 6 weeks, significant improvements were observed in measures of quality of life and social engagement. Minimal cybersickness was observed. One symptom of cybersickness, fatigue, even demonstrated a decrease over the time of the intervention. Qualitative interviews confirmed significant findings of enhanced wellbeing. While these results are promising with respect to the acceptability and feasibility of the intervention, impacts on wellbeing and quality of life should be interpreted with some caution given the small sample size and lack of a comparison group. Benham et al. (2022) demonstrated the feasibility of a different leisure-oriented VR program in a community senior center. Older adults with and without cognitive impairment completed a total of eight VR sessions across five weeks, with each session lasting 30 min. Although results were promising, again, the sample size was small, with a total of sixteen participants.

For a broader view of the literature, a relatively recent systematic review identified fifteen research articles related to the use of VR technology to promote enrichment activities among older adults. This paper defined enrichment activities as ones that "provide older people with opportunities to engage in activities that support their emotional needs and/or assist in maintaining connections with society," and the review excluded papers that focused on VR to diagnose, treat, or rehabilitate health problems. Common findings among these articles included improved mood post-VR exposure and the enjoyment by older people of diverse VR applications, including ones that focused on travel, facilitated social interactions, or fostered reminiscence about the past. However, discomfort and cybersickness were also common themes mentioned. While most studies noted that participants found their VR experience comfortable and developed either no or minimal symptoms of cybersickness, this was not always the case. Moving forward, studies examining VR as a means to help older adults engage in meaningful leisure activities should continue to evaluate both the positive impacts of these interventions and their potential negative side effects.

## 7.5 Key Initiative

Evidence to date supports the promise of XR applications, in particular VR applications, to promote leisure activities among older adults with diverse abilities. Consistent with this promise, some senior centers and assisted living facilities have already begun to offer these options to their clients, and some VR companies are marketing their products directly to older consumers and organizations that serve them. However, our understanding of the positive and negative impacts of these programs and the generalizability of research findings is limited for several reasons. First, many research studies draw their conclusions from small sample sizes. Second, most studies examine the impact of these programs without a comparison or control group. Third, more work needs to be done to better understand older adults' preferences for the content of these programs, and how to accommodate individual differences in preference. Finally, we know little about older adults' motivations to engage in these activities over the long term. Many previous investigations have examined older adults' interactions with XR within a single session. Some have extended this to a few weeks or months. Almost no studies have examined motivation to continue to engage in these activities beyond that. It is important to know how interesting and appealing these activities might be after the effect of novelty has worn off. These limitations, gaps, and unanswered questions provide an agenda for future research.

Researchers and designers continue to explore the feasibility, usability, acceptability, benefits, and potential side effects of implementing various leisure activities within the context of XR. For instance, there is limited research focusing directly on supporting creative endeavors, like XR apps designed for artistic or musical creativity. Also, while gaming is among the most popular leisure activities linked with XR, there's been scant attention given to XR games specifically tailored for older adults, taking into account the typical differences in abilities and preferences between older and younger adults. Nonetheless, we might leverage the extensive research on older adults' video game preferences for valuable insights (see Box 7.1). Ideally, a program designed to support leisure activities for older adults should offer a diverse range of choices, encompassing applications related to travel, socializing, productive engagement, and gaming. All these should be accessible through an intuitively designed interface, complete with ample training opportunities and instructional support.

**Box 7.1 Game Preferences of Aging Adults**

At this point, there is a fairly substantial literature on older adults' video game preferences (e.g., Boot et al., 2020). Fast-paced action games are some of the most popular games among younger players; however older adults tend to report a preference for games that are more intellectually stimulating, involve exploration, and are familiar to them (Fig. 7.2). Fantasy elements are typically rated as uninteresting, and older adults often report a strong aversion to violent game content. However, when considering age-related differences in game preference, it is important to avoid the ageist notion that all older adults prefer the same type of game. Like younger adults, older adults will have different preferences. In addition to preference, video games need to account for age-related changes in perception and cognition that influence the difficulty of the game. Video games that consistently challenge players at an appropriate level can create an enjoyable feeling of "flow" or "being in the zone," a concept rooted in flow theory. This state emerges when one's abilities align closely with the game's demands. However, designing a game to induce flow across the lifespan can be challenging. A game that engenders a flow state in younger adults might frustrate older players due to age-related ability changes. Video games often offer selectable difficulty levels (e.g., easy, medium, hard); however, the calibration of these levels might not consider age-related ability changes. An "easiest" option might benefit not just older adults, but novice gamers of all ages.

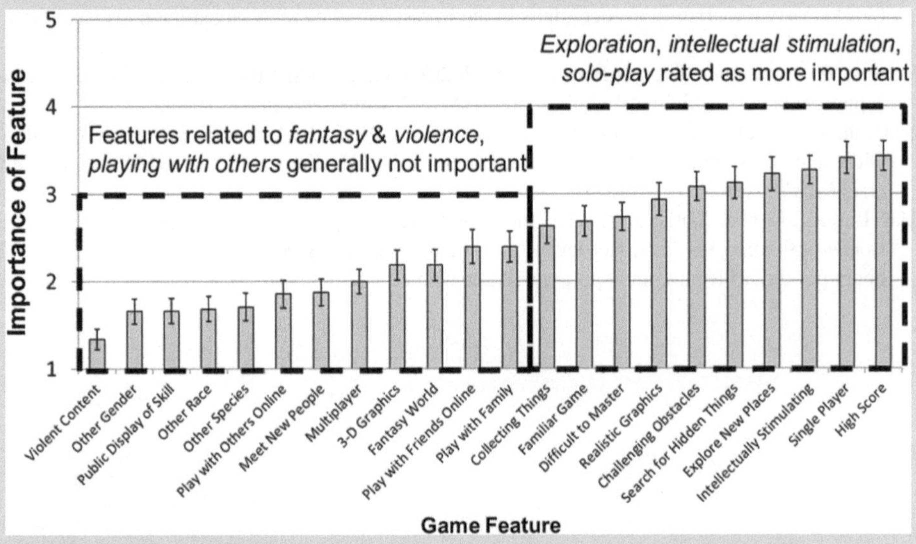


**Fig. 7.2** Older adults' rating of how important game features are (for older gamers), or would be (for older non-gamers), to their enjoyment of a video game (adapted from Blocker et al., 2014). 3 = neutral

## References

Appel, L., Appel, E., Bogler, O., Wiseman, M., Cohen, L., Ein, N., & Campos, J. L. (2020). Older adults with cognitive and/or physical impairments can benefit from immersive virtual reality experiences: A feasibility study. *Frontiers in Medicine, 6*, 329.

Benham, S., Trinh, L., Kropinski, K., & Grampurohit, N. (2022). Effects of community-based virtual reality on daily activities and quality of life. *Physical & Occupational Therapy in Geriatrics, 40*(3), 319–336.

Blocker, K. A., Wright, T. J., & Boot, W. R. (2014). Gaming preferences of aging generations. *Gerontechnology: International Journal on the Fundamental Aspects of Technology to Serve the Ageing Society, 12*(3), 174–184.

Boot, W. R., Andringa, R., Harrell, E. R., Dieciuc, M. A., & Roque, N. A. (2020). Older adults and video gaming for leisure: Lessons from the center for research and education on aging and technology enhancement (CREATE). *Gerontechnology, 19*(2).

Brimelow, R. E., Dawe, B., & Dissanayaka, N. (2020). Preliminary research: Virtual reality in residential aged care to reduce apathy and improve mood. *Cyberpsychology, Behavior, and Social Networking, 23*(3), 165–170.

Fiocco, A. J., Millett, G., D'Amico, D., Krieger, L., Sivashankar, Y., Lee, S. H., & Lachman, R. (2021). Virtual tourism for older adults living in residential care: A mixed-methods study. *PLoS ONE, 16*(5), e0250761.

Rogers, W. A., Mitzner, T. L., & Bixter, M. T. (2020). Understanding the potential of technology to support enhanced activities of daily living (EADLs). *Gerontechnology, 19*(2).

Szanton, S. L., Walker, R. K., Roberts, L., Thorpe, R. J., Wolff, J., Agree, E., & Seplaki, C. (2015). Older adults' favorite activities are resoundingly active: Findings from the NHATS study. *Geriatric Nursing, 36*(2), 131–135.

U.S. Bureau of Labor Statistics. (2001). *Men spent 5.6 hours per day in leisure and sports activities, women 4.9 hours, in 2021.* Retrieved from: https://www.bls.gov/opub/ted/2022/men-spent-5-6-hours-per-day-in-leisure-and-sports-activities-women-4-9-hours-in-2021.htm

# Other Applications

# 8

## 8.1 Introduction

As a broad overview, this book can only scratch the surface of the potential that extended reality (XR) solutions have in supporting the needs of older adults, both with and without cognitive impairments. Thus far, we have discussed how XR can support wellness, cognition, transportation, and leisure, but there are many other possibilities for the role of XR. This chapter highlights a few additional application areas in which XR has the potential to improve the health, wellbeing, quality of life, and independence of older individuals.

## 8.2 XR and Care Delivery

As previously mentioned, due to population aging, there will be more older adults than ever before in the world. Unfortunately, the number of healthcare workers with expertise in working with this demographic will significantly fall short of the required level to provide adequate care. For instance, in the United States, only 3% of licensed psychologists specialize in geropsychology, the field of psychology focused on the mental health and wellbeing of older adults (Merz et al., 2017). Yet, for a variety of reasons, many older people have a need for mental health services.

Medical extended reality (MXR), the integration of XR into the healthcare system, could help bridge the gap between patient needs and available providers and resources (Spiegel et al., 2024). For example, MXR telehealth options might offer remote access to immersive mental health services or provide automated, artificial intelligence-based counseling with a virtual agent. Related to clinical applications, virtual reality (VR) has already been used successfully in the realm of exposure therapy for treating anxiety disorders and PTSD (Rizzo et al., 2019). Similar approaches might allow older adults, especially older



W. R. Boot et al., *Extended Reality Solutions to Support Older Adults*, Synthesis Lectures on Technology and Health, https://doi.org/10.1007/978-3-031-69220-8_8

adults in remote communities, to better access mental health services. These services might not only provide benefits to older adults but could also help facilitate access to care for family caregivers who care for older adults with health and cognitive challenges.

## 8.3 XR and Pain Management

Chronic pain can be debilitating, interfere with many daily activities, and is associated with numerous poor outcomes, such as substance abuse, cognitive decline, and increased suicide risk. Over 30% of older adults experience chronic pain, and more than 10% experience high-impact chronic pain, meaning their pain levels substantially restrict daily activities (Centers for Disease Control, n.d.). A number of studies have explored VR as a tool to reduce acute pain, such as pain from burns or surgical procedures, finding this option to be safe and effective (Dreesmann et al., 2022). VR appears to have this affect through a combination of distraction, immersion, and possibly alterations in the brain's processing of pain signals.

Recently, Goudman et al. (2022) conducted a meta-analysis specifically focusing on chronic pain. This analysis, which included 41 studies, found significant evidence of VR's efficacy in reducing pain and improving functioning among chronic pain sufferers. Overall, the effects were quite large, including reductions in pain and improvements in functioning, with smaller, yet still robust, effects on mobility and functional capacity. Although this meta-analysis did not specifically focus on older adults, it supports the notion that XR solutions offer significant potential in aiding those experiencing persistent pain, affecting a range of crucial outcomes.

## 8.4 XR and Rehabilitation

Age-related brain injuries, including strokes and traumatic brain injuries (TBIs), can impair both cognitive and physical functions, often necessitating neurorehabilitation. Age is a significant risk factor for strokes, and many older adults experience TBIs due to age-related increases in the likelihood of falls and a greater vulnerability to injuries from automobile accidents. XR has the potential to make rehabilitation activities more enjoyable and engaging. XR rehabilitation programs also have the potential to be delivered remotely (Fig. 8.1).

Hao et al. (2023) recently conducted a meta-analysis of nine studies examining the impacts of virtual reality-based telerehabilitation for stroke patients. Compared to more traditional, in-person rehabilitation, remote VR-based programs produced similar effects, indicating that there is no cost to delivering care remotely. Furthermore, compared to more traditional rehabilitation exercises, XR solutions can make exercises more interesting and fun, which has the potential to increase engagement and adherence.

**Fig. 8.1** An older adult engages in remote rehabilitation exercises (Shutterstock)

## 8.5 XR and Everyday Tasks

XR, and specifically AR, has the potential to assist older adults in performing various essential daily and community living tasks. Augmented reality (AR) glasses, for example, could provide cues to an older adult with cognitive impairments, aiding in everyday activities like housekeeping and cooking, thereby enhancing their independence and quality of life. In completing complex everyday activities, AR can offer "knowledge in the world"—digital cues overlaid onto the real world that guide an individual through each step of a multi-step process. Such applications have already been successfully implemented in the industry and manufacturing sectors (Wang et al., 2016). However, to date, relatively few studies have explored XR's potential in this domain, and very few studies have involved older adult participants.

Williams et al. (2021) conducted preliminary work in this area, discovering that AR prompts could successfully guide aging adults (ages 50 + years) in performing arbitrary laboratory tasks like moving objects to different locations. Furthermore, they found that certain types of cues, such as an augmented reality "ghost hand," were more effective than others in guiding participants' actions, resulting in superior performance and participant preference. Similar support systems could assist older adults with cognitive challenges in the performance of various activities at home, such as preparing nutritious meals or completing housekeeping tasks.

## 8.6 XR and Memory Support

In the future, AR technology holds promise for supporting the prospective memory of older adults within the home environment. Prospective memory refers to remembering to complete an action in the future, such as remembering to take medication at a certain time or with a specific meal. Prospective memory, which is crucial for the execution of intended actions in the future, can often decline with age, especially among older adults with cognitive impairments.

By integrating AR into the home setting, we can provide older adults with cognitive impairments with timely reminders and notes at specific locations relevant to their daily tasks via AR glasses. For example, an AR system could overlay a virtual note next to the washer and dryer reminding an older adult with cognitive impairment that they need to do laundry soon, or project a reminder near the door to bring an umbrella if it is likely to rain that day. Similarly, a notification could appear near their medicine cabinet to prompt them to take their daily medication. Contextual, just-in-time reminders embedded in the home environment have the potential to greatly support prospective memory abilities, enhance independence, support wellbeing, and promote community living.

Other memory abilities may be supported as well. Many older adults report trouble with remembering names. AR, combined with facial recognition technologies, may help solve this problem by unobtrusively providing this information to older adults who experience this problem.

## 8.7 Summary

This chapter provided a brief overview of additional areas where XR could support older adults, both with and without cognitive impairments. As technology, including advancements in artificial intelligence and computer vision, progresses, the potential of XR solutions will only expand. We encourage readers to think creatively about other domains and tasks where AR and VR could assist older individuals.

## References

Centers for Disease Control. (n.d.). Chronic pain among adults—United States, 2019–2021. Retrieved from 10 March 2024 https://www.cdc.gov/mmwr/volumes/72/wr/mm7215a1.htm

Dreesmann, N. J., Su, H., & Thompson, H. J. (2022). A systematic review of virtual reality therapeutics for acute pain management. *Pain Management Nursing, 23*(5), 672–681.

Goudman, L., Jansen, J., Billot, M., Vets, N., De Smedt, A., Roulaud, M., & Moens, M. (2022). Virtual reality applications in chronic pain management: Systematic review and meta-analysis. *JMIR Serious Games, 10*(2), e34402.

Hao, J., Pu, Y., Chen, Z., & Siu, K. C. (2023). Effects of virtual reality-based telerehabilitation for stroke patients: A systematic review and meta-analysis of randomized controlled trials. *Journal of Stroke and Cerebrovascular Diseases, 32*(3), 106960.

Merz, C. C., Koh, D., Sakai, E. Y., Molinari, V., Karel, M. J., Moye, J., & Carpenter, B. D. (2017). The big shortage: Geropsychologists discuss facilitators and barriers to working in the field of aging. *Translational Issues in Psychological Science, 3*(4), 388–399.

Rizzo, A., Thomas Koenig, S., & Talbot, T. B. (2019). Clinical results using virtual reality. *Journal of Technology in Human Services, 37*(1), 51–74.

Spiegel, B. M., Rizzo, A., Persky, S., Liran, O., Wiederhold, B., Woods, S., & Zhang, H. (2024). What is medical extended reality? A taxonomy defining the current breadth and depth of an evolving field. *Journal of Medical Extended Reality, 1*(1), 4–12.

Wang, X., Ong, S. K., & Nee, A. Y. (2016). A comprehensive survey of augmented reality assembly research. *Advances in Manufacturing, 4*, 1–22.

Williams, T. J., Jones, S. L., Lutteroth, C., Dekoninck, E., & Boyd, H. C. (2021). Augmented reality and older adults: a comparison of prompting types. In *Proceedings of the 2021 CHI conference on human factors in computing systems* (pp. 1–13).

# Safety, Design, and Implementation Issues 9

## 9.1 Introduction

Now that we have outlined in the previous chapters specific ways that extended reality (XR) solutions can benefit older adults with and without cognitive impairments, we can return to the overall consideration of XR design factors that affect older adult populations. This chapter presents a general summary of these concerns as they apply across multiple domains of XR.

## 9.2 Fall and Injury Considerations

### 9.2.1 Injury Risk

Physical changes that occur naturally with aging, such as declines in muscle strength and flexibility, increase our risk of falling. In addition, due to decreased bone density, slower healing, and a higher risk of complications, the consequences of a fall are much more severe for older adults compared to younger ones. This makes falls one of the most salient concerns to keep in mind when encouraging older adults to adopt XR technologies. For example, VR headsets can deprive the user of crucial visual information necessary to avoid tripping hazards, and the immersive nature of VR can result in a loss of spatial awareness, making it more likely that the user will collide with physical obstacles in the environment. AR systems tend to present the user with a less obstructed view of their surroundings; however, as discussed previously (Chap. 6), even AR can pose significant fall risks if users pay too much attention to the AR elements at the expense of monitoring the physical environment.



W. R. Boot et al., *Extended Reality Solutions to Support Older Adults*, Synthesis Lectures on Technology and Health, https://doi.org/10.1007/978-3-031-69220-8_9

### 9.2.2 Injury Risk Mitigation

Careful design and the implementation of safety countermeasures can help to minimize these concerns. Many VR headsets incorporate a guardian/boundary feature that enables users to delineate a safe, obstacle-free use area. These features often require a setup process, in which the user informs the system of an area within the room where the user is free to walk and move their body without the risk of bumping into or tripping over obstacles. Once enabled, the VR system can provide a warning when the user approaches the edge of this designated space. Moreover, if this boundary is crossed, the system can switch from presenting a view of the virtual world to a view of the real world to help the user avoid obstacles in the room and navigate back to the safe area. These features make it less likely that VR users will be injured through a fall or a collision with a wall or piece of furniture. However, mobile obstacles, such as other people or pets that may wander into the safe area, can still present a hazard.

Some XR systems rely on a wired connection between the headset and a computer. This presents a tripping hazard not just for the user, but for others who may be in the room. Ceiling mounted pully systems can help with cable management, preventing the user or others from becoming entangled and falling. However, the installation of such systems may not always be practical or aesthetically pleasing for home-based use. Wireless XR headsets can eliminate such cord hazards altogether, but are often limited in their computing power due to having to pack larger amounts of electronics and a power supply into the limited space and weight requirements of a headset.

For some XR activities, and especially for individuals who are experiencing age-related changes in balance, seated XR experiences may be the most appropriate. A swivel chair, in combination with a wireless XR headset, can still allow a high degree of interactivity with the XR environment, while minimizing the risk of physical harm. Like many issues in design, there are important trade-offs to consider, since seated positions limit physical movements and may not be appropriate for some XR goals (such as certain exercise programs). Non-seated options also require careful consideration of the available space for the individual to move given their living context. A small city apartment, for example, will present challenges for safely engaging in some XR activities. Designers need to carefully weigh the risk of falls against both the product's goals and individual user needs, to ultimately reach an effective design decision. A careful, iterative, user-centered design approach can be very helpful in evaluating these factors and developing safe and effective XR applications.

## 9.3 Headset Comfort and Vision Accommodations

### 9.3.1 Headset Challenges

For all individuals, but perhaps more so for older users due to age-related physiological changes, headset comfort is an important issue. Headset weight is a concern, as well as how tight the headset must fit to be fully secure. For example, according to early reviews of the Apple Vision Pro, many early adopters felt that the headset was too heavy to wear for extended periods. It should also be recognized that many older adults wear glasses due to age-related changes in vision, which can present challenges with head-mounted displays.

### 9.3.2 Headset Solutions

As these technologies continue to evolve, headset weight will undoubtedly decrease. For now, when XR is considered as a solution to the challenges of older adults with and without cognitive impairment, maximizing headset fit and reducing the amount of time the headset is worn are two potential routes to improve comfort. In terms of visual accommodations, various solutions are available, including systems with lenses that can be adjusted further from the eyes and inserts that create additional room between the headset and face to better accommodate glasses. However, care must be taken to avoid scratching the glasses/system lenses and bending the glasses' frames due to pressure from the headset, especially as the headset is being put on and taken off. Prescription lens adapters that fit directly into the headset are also available, allowing for vision correction without requiring the user to wear glasses.

## 9.4 Cybersickness

### 9.4.1 Cybersickness Challenges

Cybersickness is a form of motion sickness that some people experience when interacting with digital environments. It is characterized by symptoms of dizziness, nausea, and headache, and is generally believed to stem from a mismatch between the visual cues of motion that a person sees and the physical motion (or lack thereof) felt by their body. The symptoms may also be affected by visual lag or stutter when motion in a VR environment is not smooth and fluid. The intensity of cybersickness symptoms can vary, influenced by an individual's susceptibility and the specific properties of the XR environment. A typical example is a driving or flight simulation, in which visual cues can indicate rapid movement and inertial forces while the body actually remains stationary. In some cases anticipation of physical forces when none actually occur can lead to postural instability, as users adjust their body in preparation for counteracting the non-existent forces. Research indicates that older adults can be more prone to such effects in simulated environments (Seifert & Schlomann, 2021).

### 9.4.2 Cybersickness Mitigation

The potential for cybersickness should be considered in the development and deployment of XR applications for older adults (Fig. 9.1). Fortunately, there are effective methods to help mitigate this risk. Mismatches between visual information and the body's perception of motion can be reduced simply by reducing the speed of available virtual movements (for example, walking rather than driving), and possibly by employing novel locomotion options such as teleportation. Advances in modern XR headsets are also reducing the lag between inputs and visual changes in the XR world, resulting in more seamless, one-to-one correspondence between user movements and corresponding view changes. However, some forms of virtual locomotion remain more likely to induce cybersickness. These include situations in which individuals use a gamepad or joystick to move themselves continuously and rapidly through the virtual environment, and situations in which

**Fig. 9.1** Cybersickness is a risk that needs careful consideration when implementing XR solutions. Symptoms may include dizziness, nausea, and headache (Shutterstock)

participants are moved through the environment by the application itself from one virtual location to another. If these locomotion options are preferred by developers for a particular reason, then shrinking the size of the visual field seen by the user while they are moving through the environment may help to reduce cybersickness.

Designers should also be aware that some individuals are more susceptible to cybersickness than others. For some applications, it may be useful to ask older adults to complete the Visually Induced Motion Sickness Susceptibility Questionnaire (VIMSSQ) before they interact with an XR environment, to determine if they are at a high risk for negative effects (Keshavarz et al., 2019). If an individual is identified as being at a greater risk for cybersickness, then the properties of the XR experience might be modified to accommodate this risk, or alternative non-XR applications might be considered. Continuous monitoring for cybersickness is recommended, especially in unsupervised, home-based XR use where unchecked symptoms such as disorientation and dizziness could increase the risk of falls. If symptoms seem to be emerging, then it is advisable to reduce or terminate exposure to the specific XR application that is causing them. For medical and psychosocial interventions, participants may first be exposed to the XR components in a controlled setting, such as a laboratory or clinic, to gauge their risk of experiencing sickness at home.

Finally, education is a vital aspect of safeguarding older adults from the adverse effects of cybersickness. Users should be informed about potential symptoms and instructed to stop the XR experience if they detect any early signs to prevent subsequent exacerbation. While researchers have found that XR applications are generally well tolerated and safe for use among older adults, the potential negative effects of cybersickness should not be ignored and must be navigated systematically to help ensure user comfort and safety. In our own laboratories, we have observed few significant cases of cybersickness when using modern VR headsets and a careful selection of locomotion options. By avoiding the motion mismatches that appear to be the root cause of cybersickness, designers can help to ensure that these experiences are safe and enjoyable for users.

## 9.5 Enhancing Presence

### 9.5.1 What is Presence?

Presence is a key concept in XR experiences, predominantly discussed in the context of VR. It is described as "the sense of being there" and is considered one of the defining characteristics of the technology. In essence, presence refers to the perception that one is genuinely in the virtual world rather than in the real one. This concept is vital because a high degree of presence has been correlated both with enjoyment of the environment and with the success of VR applications in achieving various goals, such as physical rehabilitation or learning outcomes. A growing body of research also suggests that higher degrees of presence are correlated with reduced cybersickness (Weech et al., 2019).

### 9.5.2 Factors that Shape Presence

Properties of the VR experience, as well as characteristics of the individual, both play roles in shaping a user's sense of presence (Felton & Jackson, 2022). One fundamental aspect is visual immersion, which denotes the extent to which a VR display encompasses the user's view and blocks out the physical world. A seamless visual experience (e.g., minimal lag, adequate refresh rate) combined with binaural auditory cues can amplify feelings of presence. Tactile sensations or haptic cues can further augment the feeling of physically interacting with the virtual environment. Collectively, these elements coalesce to form a compelling and immersive VR setting. Gustatory and olfactory cues (taste and smell) remain largely untapped as contributors to VR presence due to the technological hurdles associated with stimulating these senses.

While some cues can enhance presence, others can detract from it. Real-world sensory distractions external to the VR environment can undermine presence. The Magnet Model of Spatial Presence conceptualizes this as a push and pull relationship between cues that draw attention to either the virtual world or the real one (Mitzner et al., 2021). The contextual salience and intensity of these cues play a crucial role. For example, while experiencing a virtual nature scene, a blaring siren from an emergency vehicle in the real world is likely to pull the user's attention, making them feel less present in the virtual world. If, instead of a virtual nature scene, the user was exploring the streets of a virtual city, then the same siren would not seem so out of place and thus would less likely disrupt their sense of presence. Other sensory cues, such as the weight of the VR headset or a physical discomfort, can also divert focus from the virtual experience. Overall, therefore, facilitating presence involves considering the dynamics between virtual and real-world cues that the user is likely to experience. Presence will benefit from strategies that either eliminate cues originating from the real world, or decrease these cues' incommensurability with the VR content.

Beyond the properties of the XR experience, individual traits also influence the degree of presence that is felt. People differ in the personality trait of "absorption," reflecting their ability to fully concentrate on a task. Some studies have found that those with higher absorption traits often feel a greater sense of presence in VR environments. However, many other factors of personality and individual difference have been linked to experiences of presence, and to some extent these experiences may also depend on how well the VR content relates to one's personal background and interests.

### 9.5.3 Presence and Age

Considering the focus of the current book, it is important to discuss how experiences of VR presence might fluctuate with age. To feel embedded within the VR environment, one must form a mental representation of the virtual world, relying heavily on spatial cognition. Attentional abilities are pivotal too, in that they help us tune out distractions. Consequently, several perceptual and cognitive skills that are known to decline with age are implicated in the experience of VR presence, raising the possibility that older adults might face particular challenges in establishing and sustaining presence compared to their younger counterparts. While this is an interesting theory, current empirical research findings do not support it. Dilanchian et al. (2021), for example, asked younger and older adult participants to interact with various VR applications, and assessed experiences of presence through post-exposure surveys. Intriguingly, older adults tended to report a significantly *greater* sense of presence compared to younger ones. Dilanchian and colleagues speculated that virtual environments might be more novel for the older adult participants, given that younger adults are more likely to have prior experience in interactive gaming. This novelty could have the effect of enhanced attention to the virtual world, thus bolstering experiences of presence among the older adult participants, at least in the short term.

Mitzner et al. (2021) also evaluated potential age differences in the experience of presence using continuous measurements during VR exposure. Younger and older adult participants adjusted a slider depending on whether they felt more present in the real world or in the virtual world. Both age groups quickly felt a sense of virtual presence, and this high degree of presence in the virtual world did not differ as a function of age. Over time, however, the younger adults tended to shift back and forth from VR presence to real-world presence, while the older adults had notably fewer breaks in their experience of virtual presence. Evaluating the overall research in this area, a recent meta-analysis by Martingano et al. (2023) found that there was no consistent evidence for a relationship between age and the ability to experience VR presence. Thus, there appears to be minimal cause for concern that VR applications would need to be specifically adjusted to promote greater presence for older adults.

In the larger picture, presence remains an important consideration in all types of VR development due to its crucial impact on the enjoyability and effectiveness of the technology. For the most part, designers creating XR experiences for older adults can rely on the standard facets known to improve the sense of presence for users, encompassing both hardware and software components (Souza et al., 2021). It is also recommended, however, that consideration should be given to developing content that is likely to be of interest to the older adult population (including specific sub-population groups), since the salience of the content to users' personal interests can significantly contribute to the maximization of presence.

## 9.6 Other Considerations

### 9.6.1 Intervention Implementation

Premature *implementation* of technology solutions can lead not only to ineffective applications that are unlikely to be adopted by users, but also to applications that may have potentially dangerous consequences. Before implementation in communities, a rigorous, multi-stage evaluation process is recommended to review new XR products targeted toward older adults. Many applications never reach the level of wide-scale dissemination and adoption due to a failure to consider the potential barriers that may be faced in getting the technology to users. The NIH Stage Model for Behavioral Intervention Development provides a useful roadmap for developing safe, effective, and scalable psychosocial interventions (Onken, 2022), which may usefully be adopted by XR designers. Within this framework, early pilot studies examine issues of feasibility, acceptability, and initial efficacy under relatively controlled conditions. The application is then refined, and its efficacy is further confirmed through larger scale randomized controlled trials to better characterize benefits as well as potential negative consequences. Later stages involve studies to understand the effectiveness of the application beyond controlled research settings, and to develop strategies to effectively implement and promote it within the community. This framework highlights that the development and testing of successful products must involve a careful, systematic, and multi-stage process, involving several types of studies with different aims (e.g., determining feasibility vs. efficacy vs. disseminability). While requiring a significant investment in time and resources, such an approach can serve to maximize the benefits of an XR product while minimizing the likelihood of adverse consequences.

### 9.6.2 Training and Instruction

It is also very important to provide *training and instructional support* for XR products targeted toward older adults. In addition to the physical and mental challenges that naturally develop with age, older adults are less likely than younger adults to have experience and proficiency with recently emerging technologies. Researchers have found that, on average, an older adult may take nearly twice as long to learn a new software application compared to a younger adult, if they have little previous experience with similar applications (Charness et al., 2001). These challenges are compounded by the fact that developers are often younger adults themselves, and may not adequately account for age-related differences in abilities and knowledge in their product designs. Effective training and instructional support can mitigate these problems and greatly enhance older adults' ability to comfortably interact with XR technologies, thus maximizing the benefits received. An iterative, user-centered design approach has been central to much of the discussion in this book.

The development of training protocols and instructional support—such as manuals, tip-cards, tutorials, and online help systems—is no exception. Before implementing a new XR product, such materials should be carefully developed and tested with representative older users, with subsequent revisions based on the user feedback. Designers should recognize that learning rates are likely to be slower for older adult users, and should follow established guidelines for the creation of training and instructional support materials targeted toward the older adult population (Czaja & Sharit, 2012).

## 9.7 Summary

This chapter provided an in-depth discussion of age-related cautions and considerations in the design of XR technologies. Of particular concern are safety issues related to falls and cybersickness. These potential risks and their unique dimensions for older adults need to be understood and thoroughly addressed before an XR product is commercially distributed. Issues related to enhancing the sense of presence in VR environments, conducting adequate product testing with older adult users, and providing sufficient training and instructional materials were also discussed. Designers who are not older adults themselves may have difficulties intuitively recognizing the needs and perspectives of this population, and thus a systematic, multi-step, user-centered development agenda is vital. Adopting this approach can lead to the creation of XR solutions that are safe, useful to, and usable by diverse older adults in the community.

## References

Charness, N., Kelley, C. L., Bosman, E. A., & Mottram, M. (2001). Word-processing training and retraining: Effects of adult age, experience, and interface. *Psychology and Aging, 16*(1), 110–127.

Czaja, S. J., & Sharit, J. (2012). *Designing training and instructional programs for older adults*. CRC Press.

Dilanchian, A. T., Andringa, R., & Boot, W. R. (2021). A pilot study exploring age differences in presence, workload, and cybersickness in the experience of immersive virtual reality environments. *Frontiers in Virtual Reality, 2*, 736793.

Felton, W. M., & Jackson, R. E. (2022). Presence: A review. *International Journal of Human-Computer Interaction, 38*(1), 1–18.

Keshavarz, B., Saryazdi, R., Campos, J. L., & Golding, J. F. (2019). Introducing the VIMSSQ: Measuring susceptibility to visually induced motion sickness. In *Proceedings of the human factors and ergonomics society annual meeting* (Vol. 63, No. 1, pp. 2267–2271). Sage.

Martingano, A. J., Duane, J. N., Brown, E., & Persky, S. (2023). Demographic differences in presence across seven studies. *Virtual Reality*, 1–17.

Mitzner, T. L., McGlynn, S. A., & Rogers, W. A. (2021). Understanding spatial presence formation and maintenance in virtual reality for younger and older adults. *Gerontechnology, 20*(2).

Onken, L. (2022). Implementation science at the National Institute on Aging: The principles of it. *Public Policy & Aging Report, 32*(1), 39–41.

Seifert, A., & Schlomann, A. (2021). The use of virtual and augmented reality by older adults: Potentials and challenges. *Frontiers in Virtual Reality, 2*, 51. https://doi.org/10.3389/frvir.2021.639718

Souza, V., Maciel, A., Nedel, L., & Kopper, R. (2021). Measuring presence in virtual environments: A survey. *ACM Computing Surveys (CSUR), 54*(8), 1–37.

Weech, S., Kenny, S., & Barnett-Cowan, M. (2019). Presence and cybersickness in virtual reality are negatively related: A review. *Frontiers in Psychology, 10*, 158.

# 10 Concluding Remarks

## 10.1 Introduction

Within this book, we have outlined the exciting promise of extended reality (XR) solutions to support the needs of older adults, both with and without cognitive impairments. These solutions encompass both virtual reality (VR) and augmented reality (AR) technologies, which hold the potential to help older adults stay connected with friends and family, engage in activities that support healthy lifestyles and wellness, transport themselves from one location to another safely and confidently, maintain cognitive fitness, overcome cognitive challenges, and engage in meaningful leisure opportunities. The goal of this book is to inspire readers to consider additional solutions both within and beyond these activity domains by highlighting the needs of older adults and various possibilities for support. This chapter concludes with important issues to consider as these research and design efforts move forward. These issues include limitations in our understanding of the effectiveness of some XR solutions, the challenges of applying design guidelines to new technologies that may be fundamentally different from the technologies that preceded them, the difficulty of defining a research agenda in the midst of rapid technological changes, and the consideration of ethical issues involved when technology has the potential to replace human contact.

## 10.2 Limitations in Design Guidelines

New technologies can generally be categorized into two groups: incrementally new technologies and radically new technologies. For instance, a computer tablet might be considered an incrementally new technology because it retains many functions of a traditional computer. Apart from its portability and the input method (e.g., virtual rather than



W. R. Boot et al., *Extended Reality Solutions to Support Older Adults*, Synthesis Lectures on Technology and Health, https://doi.org/10.1007/978-3-031-69220-8_10

physical keyboard), it operates in much the same way. Thus, one's "mental model" of how to interact with a computer can largely be repurposed, with minor adjustments, for successful tablet use. In the design of incrementally new technologies, a wide range of design challenges can often be anticipated based on existing technologies and guidelines for designing for older adults.

Conversely, autonomous vehicles can be viewed as a radically new technology. The automation fundamentally alters the nature and requirements of the driving task, transitioning the role of the driver from active control to passive monitoring. This presents entirely new challenges, such as managing the transition of control from automated driving back to the human driver when the automation fails. Thus, much greater attention should be devoted to understanding issues of usability and age-related differences in performance given the new demands and problems that can arise that may not always be easily anticipated. On the spectrum between incrementally new to radically new technologies, XR solutions often fall more toward the radically new end of the continuum.

While existing design guidelines for designing for older adults provide useful guidance (e.g., Czaja et al., 2019), much more research is needed to develop more comprehensive, empirically based guidelines for how to design XR solutions specifically for diverse groups of older users. This is a critical research agenda. Currently, the pool of literature available to inform the best ways to design XR technologies and applications for diverse groups of older users is limited. As discussed in Chap. 9, issues of safety and comfort are paramount.

## 10.3 Limitations of Efficacy Evidence

The XR solutions discussed in this book are intended to improve the health, wellbeing, quality of life, and social connectedness of older people. Much of the data reviewed in each chapter supports the feasibility, usability, and acceptability of these approaches. Some also provide initial indications of efficacy, that is, the ability of the XR solution to improve important outcomes related to older adults' health and wellbeing. However, many of these trials feature small sample sizes and either no control group or a control group that does not adequately account for potential experimental confounds. For example, a "passive" or "sit and wait" control group that engages in no alternative activity is unlikely to rule out placebo effects (Boot et al., 2013). The design of these studies limits our understanding of the effectiveness of these approaches. Caution is warranted in interpreting the impact of these XR solutions on the lives of older adults until more, larger, and more rigorous studies are conducted. These types of large-scale studies will help prevent the rollout of ineffective XR applications and interventions.

Within these studies, it is sometimes difficult to understand *why* a particular XR solution might work. For example, if an XR application reduces feelings of loneliness, how does it do so? Is the reduction in loneliness, for example, caused by the XR application's ability to increase an individual's perceived social support? Such a variable, whose change is a precursor to a change in the targeted outcome, is often referred to as a mediator. By better understanding mediators that act between the intervention and the desired outcome, there is an opportunity to adapt the intervention to target this mediator better, resulting in even greater reductions in loneliness, or change in other outcomes of interest. Small-scale studies that only assess outcome measures without measuring potential mediator variables limit our ability to develop more successful XR solutions. Mediators can also be better understood by varying the experience of the XR group compared to the control group systematically in order to identify one or more "active ingredients" within the XR application responsible for better outcomes.

Current limitations in the evidence base in support of XR solutions should be considered in advance of their implementation. However, the need for larger, more powerful studies that also focus on potential mediators of improvement provides a clear research agenda over the next few years.

## 10.4 The Rapid Pace of Innovation

Writing about technology, especially in a domain as dynamic as XR, is inherently challenging due to its rapid evolution. It is important to recognize that this book was completed in 2024. In just a few years, the technological landscape will almost certainly be very different from what it was when this book was published. New tools and systems are constantly emerging (and disappearing); for instance, the release of Apple's Vision Pro, a sophisticated AR system, underscores the pace at which innovation unfolds, offering enhanced potential to support older adults. However, although it is clearly an advanced piece of technology, the future of the Vision Pro remains uncertain. Additionally, in the realm of artificial intelligence, the swift rise of sophisticated large language models (LLMs) signals a promising future where they may be seamlessly incorporated into XR platforms, enabling richer, adaptive, and highly personalized XR experiences. While these rapid advancements in LLMs hold significant promise, they also present numerous challenges and raise many important unanswered questions, including complex ethical ones. There will be no shortage of design problems to address or applications to explore with respect to improving the lives of older adults in the future.

With every technological advancement comes a fresh set of design and implementation challenges. An important goal is to understand that, while specific devices and tools might change, the foundational principles of assessing and implementing XR technologies, as outlined in this book, remain consistent. Focusing research on broader principles, rather than the intricacies of a specific device, can help ensure that research findings and design recommendations remain relevant and applicable even as the technology landscape shifts.

## 10.5 Ethical Considerations

Finally, when considering XR solutions for delivering care and supporting social engagement, it is crucial not to overlook potential ethical implications, especially with respect to XR applications that have the potential to replace genuine human contact. Relying heavily on XR in these and other applications could lead to reduced face-to-face interactions, which might negatively affect individuals' emotional wellbeing and interpersonal skills. Rather than reduce loneliness and increase social connectivity, it is possible that these technologies, at least for some individuals, could paradoxically be more isolating. In the development and deployment of these technologies, ideally, these technologies will be used to supplement rather than replace face-to-face interactions. Otherwise, they might provide support when other, more personal forms of support are not possible. Developers should consider these balances and the pros and cons of different approaches to improve the lives of older adults. Issues of preferences and autonomy should be carefully considered as well.

## 10.6 Summary

In this book, we explored the potential of VR and AR technologies in various domains, from supporting cognitive health to enhancing social engagement (Fig. 10.1). These cutting-edge tools offer the promise of enhanced connectivity, wellness, and diverse leisure opportunities for our aging population. Yet, with innovation come new challenges: the fast-paced evolution of technology, the gaps in evidence-based guidelines tailored for older XR users, and the uncertainty in the long-term effectiveness of these solutions. Notably, as XR technologies advance rapidly, maintaining the relevance of guidelines and understanding their application will be crucial. Furthermore, while there is growing evidence pointing to the benefits of XR for older adults, larger and more rigorous studies are needed to understand the full spectrum of impacts and to ensure effective outcomes. Amidst this wave of technological growth, ethical considerations are paramount. The appeal of XR should not overshadow the value of genuine human interactions, and its use should aim to enhance, not replace, personal connections. Developers must tread thoughtfully, considering both the immense potential and the ethical implications of XR solutions. As we look ahead, we hope the principles and insights presented in this book will serve as a guide for future XR application and design, emphasizing the importance of human-centered design and ethical considerations in the rapidly evolving world of XR.

**Fig. 10.1** Well-designed XR solutions have immense potential to support activities that are important to older people, offering them innovative solutions for improving health, wellbeing, quality of life, and social connectivity (Shutterstock)

## References


Boot, W. R., Simons, D. J., Stothart, C., & Stutts, C. (2013). The pervasive problem with placebos in psychology: Why active control groups are not sufficient to rule out placebo effects. *Perspectives on Psychological Science, 8*(4), 445–454.

Czaja S. J., Boot W. R., Charness N., & Rogers W. A. (2019). *Designing for older adults: Principles and creative human factors approaches* (3rd ed.). Taylor and Francis.

[illegible]

## References

[illegible]